A BIOLOGIST ABROAD

A Biologist ABROAD

Rory Putman

drawings by Catherine Putman

Whittles Publishing

Published by
Whittles Publishing,
Dunbeath,
Caithness KW6 6EG,
Scotland, UK

www.whittlespublishing.com

© 2021 Rory Putman

ISBN 978-184995-471-6

Printed in the UK by Short Run Press Ltd.

CONTENTS

BOOK 3
Indonesian adventure

PREFACE AND ACKNOWLEDGEMENTS

These tales are drawn from a number of overseas expeditions I have made over the years – one from each of the first four decades of my professional life – spanning the (sub-)Arctic, Africa and Asia. They are inevitably coloured by the perspective of a keen naturalist, but are primarily drawn from the diaries of a working biologist and so, as well as recording my impressions of the countries I visited and their people, they include some biological reflections – some of which perhaps did not make it to the formal pages of the academic literature.

Sharing my experiences in Iceland in the late 1960s were Chris Bulstrode, Simon Corbett, Adrian French, Harry Machin, Tony Taylor and Chris West; to all I extend my thanks for making their efforts in trying to deliver our objectives.

In East Africa, Lesley and Nigel Lloyd were outstanding hosts who were exceptionally generous with their time and unstinted hospitality. For giving up their time to allow us insights into their ongoing research, I thank also David and Jeanette Bygott, Mark Stanley-Price and Jeff Lewis, as well as many others who do not appear in these accounts.

I thank the British Council for inviting me to travel to northern Nigeria on its behalf, and extend a huge thanks to my companions in Nigeria itself: Dr Gadzama, senior livestock officer, and Alhaji Zanna Talha. I am grateful for the long-suffering and patient support offered by Bob Steedman, the British Council's representative in Kano, and in Maiduguri I was shown tremendous kindness by Ralph and Rosemary

Crates (who also nursed me through a period of illness self-inflicted as the result of my own carelessness and inexperience).

In Indonesia I thank my students and friends, the ever-interchangeable Martin Hedges and Simon Tyson.

I must emphasise here that the book offers my own personal impressions of these different countries at the time of my visits – in some cases 30 or 40 years ago. There may be many inaccuracies and half-truths, many misconceptions; further, all the countries described have undoubtedly changed greatly from the times in which I was their guest. Nevertheless I hope I will be forgiven any inaccuracies, any unintended slights. For this book is indeed intended rather as a homage to some of the beautiful countries in which I have been privileged to be made welcome, and the wonderful people who offered me that welcome.

In the UK, in 2019, I thank Keith Whittles of Whittles Publishing for agreeing to publish what, for me, is a major departure from my normal writing output, and owe enormous thanks and admiration to Caroline Petherick, my editor (www.the-wordsmith.co.uk), for her thoughtful and sensitive suggestions for improving the text. Over the years I have worked with a good many 'wordsmiths' and she ranks right up there.

BOOK 1

Letters from Iceland

A note on Icelandic pronunciation

The Icelandic alphabet has a number of letters not current in English. Two of their forms appear in the text; where found, they should be pronounced:

Þ : 'th' as in 'thing': Þingvellir = Thingvellir

ð : in the middle of a word, 'th' as in 'this'; thus Goðafoss = Gothafoss
: at the end of a word, 'd'.

NB the letter 'j' is pronounced as an English 'y', so Þjórsárver is thus rendered 'Thyorsarver'. The principal vowel sounds to note in this book are 'ö' = ur (as the German 'ö'), and 'æ' = I. For more on Icelandic pronunciation, see, eg, https://ielanguages.com/icelandic-pronunciation.html.

PROLOGUE

Gollum shivered. For some time now he had been sitting patiently on the braided kettle handle that passed for a perch and now he was getting bored. He shuffled sideways and looked around him with his usual defiance. We'd been lucky with this one: a scrawny bedraggled ragbag of feathers when first he was brought to us – now a malevolent dandy, sleek and suave right down to the stubby tip of his six-month-old tail. Watchfully he sidled along his kettle handle and dropped unnoticed to the floor.

I grinned across at Chris – for the young magpie was his triumph as much as mine. Chris's face was unusually serious, composed in a mask that barely concealed his delight at being on the right side of an interview table for once. He was enjoying himself: but no more so than the young man, Adrian French, who faced us across my coffee-stained desk, remarkably self-assured and obviously at ease, quietly and firmly trying to convince us that we wanted to take him on. But did we?

European Conservation Year, and four of us were planning an expedition to central Iceland, to study the wild geese that bred in the desolate central highlands: Chris Bulstrode, now doing his best to act the serious interviewer – a disreputable young medical student in actuality – was to lead the trip, accompanied by myself and Simon Corbett – committed behaviourists – and Chris, 'Root', West, the practical, pragmatic, unrufflable Root – an embryo engineer. Already the four of us were on our wild goose chase. (We were quite used to that gibe by now, indeed even quite proud of it: it had enabled us to winkle out a not inconsiderable amount of money from a trust fund dedicated to the support of crack-brained schemes. Our literal wild goose chase had seemed an ideal candidate for assistance). Did we really need another expedition member? Come to think of it, I wasn't really even sure how I'd got involved in it in the first place. I suppose Bulstrode's infectious and indomitable enthusiasm had caught me up, but however it had been, Chris and I had spent a good deal of the last few months planning and organising our trip.

The pinkfooted goose – a small and attractive member of the tribe of grey geese – breeds in only two locations in the world: Iceland, and Spitsbergen and north Norway. Yet the two populations are geographically quite distinct, with the Spitsbergen/Norway birds wintering south in Denmark and west Germany, while the Icelandic pinkfeet winter in Scotland. Our interest? To find out as much as we could about the Icelandic race and see whether or not it might constitute a distinct subspecies, and to evaluate the status of the breeding population of this Icelandic form.

To this end we planned parallel visits to both Iceland and Spitsbergen, making notes on the appearance and behaviour of the two races and obtaining blood samples from the birds to undertake electrophoretic analyses of blood serum proteins (a standard way of assessing genetic relationships between species in the days long before techniques were developed for the direct analysis of DNA).

This year we were making our maiden run to Iceland. It all sounded very reasonable on paper, but none of us really knew what we were up against: neither Chris nor I had ever been to Iceland before, and although we had sought sage counsel from many who had, our information was at best second-hand. For most of the 13 weeks of the Icelandic summer we were to be at least 25 gruelling miles from the nearest habitation and around 200 miles from any town.

'It's not going to be a rest cure, you know – no glorified holiday with the added thrill of being a scientific expedition on paper. It's going to be damned hard work, under pretty terrible conditions – why are you so keen to come?' Bulstrode again; he had reckoned we needed to take a couple more stalwarts along, if only to build up numbers for catching the geese. I wasn't so sure: the four of us in it so far were close friends; wouldn't it just be inviting trouble to take on extras, comparative strangers? What if we didn't get on, what if they couldn't take the pace? There was going to be no getting away from each other up there in the tundra; for better or worse we would all have to live right on top of each other.

'I'm not frightened of hard work. I'm a pretty fair birdwatcher and I think I can be useful to you. Besides – aaaargh! What the …?'

Discomposed for the first time since he'd stepped into the room, French clutched his foot. Gollum of course; he'd slid quietly across the floor unobserved. From his vantage point on the kettle he'd suddenly registered that the young man was wearing open-toed sandals, and if there was one thing Gollum enjoyed above almost anything it was driving his dagger-like bill at bare toes. He squawked and disappeared under the bed. But it was too late: the serious expression on Bulstrode's face slipped and was gone for good. Adrian French would join us, and would prove invaluable.

I pulled myself out of my chair. 'Let's go down for a beer on it, then.' Gollum reappeared as if by magic and settled into the crook of the chairback as I pulled the door to behind us.

In the event we numbered seven, for we also welcomed aboard Tony Taylor – glad of his experience, for Tony was a fine ornithologist and had worked in central Iceland before – and Harry Machin, as expedition artist. So there we were: two psychology students, two medics, two engineering students and an artist, a strange ménage for 13 weeks in the middle of Iceland – but it was one artist and six zoologists that were to return.

One

AS FAR AS FOSSRÓFULÆKUR

I suppose the magnitude of it all only really struck home as we stood on the docks at Leith, watching our ton and a half of food and equipment lowered onto the MS *Gullfoss*. Here we all were – not one of us over 20 years of age, having spent some £1,300 of other people's money,[1] ready for … what?

Tony and Harry were flying out to join the party in Reykjavík later; the other five of us had travelled up from London the previous day to supervise the loading of our 18 crates, and to accompany them to Iceland by sea. And a pretty hectic trip it had been so far, with only 36 hours of the expedition elapsed. Last-minute packing had resulted in another two crates needing shipment and, one way or another, we had brought them up with us from Oxford on the train. Somehow, from the looks of the regular straphangers as the crates lumbered twice around the Circle Line (we'd forgotten to get off, first time around), I got the strong suspicion that the London Underground is not supposed to cater for passengers lugging 75-kilogram crates, each a cubic metre … it had certainly proved challenging lifting the unwieldy cases over the entry barriers. But eventually all was stowed, and at King's Cross we crammed ourselves into the night train to Edinburgh. Since we had not extended to the extravagance of booking sleepers, the night was destined to be spent in a mêlée of intertwined heads and legs as everyone tried to get length enough

---

1 Equivalent to perhaps £21,000 at 2019 values.

to sleep in our overcrowded single compartment. Yet, every time I woke through the disturbed night, the carriage seemed emptier, until eventually Adrian and I woke up alone, the others having gone in search of more commodious quarters.

Edinburgh at half past six of a dull morning, and we were forced to face up to the problem of how to transport our 'excess baggage' the two miles to Leith and to the docks. In the event, the matter resolved, we careered through the early-morning streets with our enormous crates perched upon a four-wheel flatbed trolley 'borrowed' for the duration from British Rail. Amid the startled glances of Edinburgh's early risers, we negotiated traffic lights and roundabouts with equal abandon and finally rattled over the Cassey blocks to No. 8 shed at Leith harbour.

A day negotiating with bemused but relatively helpful Customs officials, and back to the city centre for another disturbed night. Ejected from the station waiting room, we slumbered peacefully on benches in Waverley Park, escaping at the crack of dawn to lie in a tidy row of quilted sleeping bags on the pavement outside, basking in the watery sunshine and waiting for breakfast. And at last, no doubt to the patient relief of the Edinburgh populace and the local police force, at last we were standing with the dock supervisor outside No.8 shed, drinking strong, black dock tea, liberally laced with a 12-year old single malt, watching the last of our crates disappear into the MS *Gullfoss*.

The Icelandic Steam Shipping Company –Eimskip for short, and which at the time of our travels was part of DFDS – used to run two boats from Leith to Reykjavík, the *Gullfoss* and the *Bruarfoss*, and two, the *Goðafoss* and the *Dettifoss*, from Hull. In 1937, W. H. Auden and Louis MacNeice wrote in their wonderful *Letters from Iceland*: 'As far as the secondclass accommodation goes it is better on the Hull boats and best on the Dettifoss. The fare from Hull to anywhere in Iceland is £4. 10s plus 5 kronur a day for food.'

But by the 1960s of that mighty fleet only the *Gullfoss* remained: a venerable but still respectable lady. We screwed the top back on the whisky bottle (Adrian had already proved his worth: his girlfriend turned out to be the daughter of one of the big distillery families!) and followed our crates on board.

I cannot in all honesty recommend travelling steerage. But then

I have always had a strange sympathy for the humble sardine, and a bunk in a dark locker next to the engine room – the intolerable heat of whose boilers was still further increased by the sheer mass of humanity crammed together as tightly as legality permitted – may be someone else's idea of heaven. Who am I to judge?

And it was cheap – a cruel irony to reflect, 50 years later, that refugees from civil war in eastern Europe and Asia might pay their life's savings to people smugglers for access to conditions probably not dissimilar and far less safe. And there were compensations. From the minute we stepped on board, hospitality was in true Scandinavian style. Steerage passengers were given the run of the dining room after the first-class clientele had eaten their fill – to pick up the crumbs from the rich man's table? And what crumbs they were! Our first-class hosts would have made hardly any impression on the succulent cold board that greeted most of the meals. To five, healthy, seasick-free appetites, this was the life. Who were we to turn up our unaristocratic noses at raw fish, at *hákarl* (half-rotten shark) or *hangikjöt* (smoked mutton)? It might not have been the Water Rat's coldhamcoldtonguecoldbeefpickledgherkinssalad, but we revelled in it.

For three halcyon days we basked in unaccustomed gastronomic luxury – daylight hours spent being buffeted about the deck by the strong winds which followed us north, glorying in the fresh weather and the abundance of ocean birds, and evenings spent nursing duty-free whiskies, yarning till the small hours. It was a magnificent, carefree voyage – the *Gullfoss* believed in service. I even got quite used to sleeping in that hot and airless locker.

And so we plodded north. The fulmar petrels which had been following our wake thinned out to the occasional dark-phase visitor. For a time the shags continued their time-honoured variation of Russian roulette: diving under the bow of the ship and flirting daringly with the blades of the propellers as they passed underneath, before emerging with a shake of the feathers in the creamy wake, to fly back forrard for another go. Then they too were gone, and only the skuas and an occasional squadron of storm petrels escorted us. The Arctic skuas had at first aroused keen interest – the first, a dark, evil-looking harpy, had joined us just beyond the Orkneys – but already we were becoming accustomed to their presence, and only the storm petrels and a rare gull-billed tern occasioned more than a second glance.

Then the fulmars joined us again, and we knew we must be approaching land; these ocean-going petrels, despite their pelagic habits, never stray far from the cliffs on which they nest. Then, late in the evening, 29 June it was, we caught our first glimpse of land: an uninterrupted horizon of ice – Vatnajökull (Water Glacier), Iceland's biggest icefield — in those days, before global warming, covering some one-third of the southern part of the country. No wonder that the first visitors called this place Iceland, for this must have been their first view too: a vast, wild, yet curiously beautiful expanse of solid ice. And sure enough, in the grey dawn of the following morning we took our places in the *Gullfoss*'s lower lounge for Immigration.

Had we brought any undeclared fishing rods? Incredulity when we affirmed that we had not – why else did the English visit Iceland if not to fish? A quick flick through the file of undesirable aliens (why do they always do that when they come to *my* passport photograph?) and then we were through.

Now the amazing thing about Reykjavík harbour in those far-off days devoid of security alarms was that there appeared to be no official buildings. No Customs, nothing. It seemed that one could have stepped off the boat, traversed the mere 30 feet of barrier-less concrete to the main road, and climbed onto the nearest bus. Our packing cases were already beginning to pile up on the side, but no one seemed in the slightest bit interested in them, or us. We really felt as if we ought to be doing something. So we wandered across to a prefabricated site hut standing at the end of the wharf and tried looking helpless. It might have been easier if we hadn't bothered, for it eventually turned out – to our horror – that we couldn't move any of our somewhat specialised equipment without a letter from the appropriate authority granting us permission to undertake scientific research within the country. And now we started looking helpless in earnest, for we had been trying to obtain such a letter for weeks. We had had informal permission, but nothing official. Why does this never happen to Attenborough?

Anyone who has ever tangled with red tape in foreign countries, being shuttled from office to office, from bored clerk to bored receptionist, from paper form to paper form for days on end – anyone who has known that will appreciate the sinking feeling that gripped us in the pit of the stomach. Needlessly, as it turned out, for the Icelanders are made of sterner stuff. Before we had really begun to realise what was happening, Chris and Tony (who'd turned up from nowhere) were whisked away by one of the Customs officials – in his own car – to visit the appropriate personages, and returned brandishing the required authorisations. The Customs officer grinned happily at a job well done and returned to his makeshift hut. And while Chris and Tony headed off again to sort out the arrangements for our transport up country, the rest of us wandered off to explore.

Iceland is a remarkable country: a fine, unspoilt land of proud, free people. Much of the country is covered by permanent ice – or was in those days; most of the rest only clears for a few short weeks in the summer. Of a population at that time of some 200,000, about 80,000 actually lived within the capital at Reykjavík. There was, and remains, only one other town of any size: Akureyri, right up in the far north of the country. For the rest, the people lived in scattered hamlets around the coastline, where, in a ribbon perhaps five to ten miles wide at the most, the alluvial plains are sufficiently fertile to permit subsistence agriculture. A single road sweeps round the coast, but in the 1960s the tarmac ran out a brief hundred metres outside Reykjavík, and many of the holdings could not in those days be reached overland, particularly during winter. Instead, they clustered around the steep fjords, their only contact with the outside world through the irregular visits of a small steamer which in the ice-free summer months used to ply its way steadily anticlockwise around the island. The steamer service has now gone – replaced by a shuttle service of light aircraft; visits are more frequent and can be continued through most of the winter. The isolation is not so great, but the people remain the same. They are a rugged, delightful race of individualists: proud, practical – and tremendously hospitable.

One other road thrusts its way up from Reykjavík, towards the centre of the island. This is the tourist route which runs as far as Geysir and the dramatic waterfall of Gullfoss; in those days it petered out there. A route of sorts continued up across the central massif to Akureyri, but it was only passable in summer, when the farmers run their sheep out to the central meadowlands behind the retreating snow. Indeed a 'route' is all it was, for beyond Gullfoss it was merely a suggestion sketched in on the map across the gravelly moraines of the central plateau; this was the route we must ourselves take in due course.

No one lives far from open countryside in Iceland: even in Reykjavík, the townspeople are of a similar mould to the country-folk – in most cases no more than a generation removed. The whole pragmatic, easy-going 'can do' philosophy of all the Icelanders we met sharply reflects the rigours of exacting a living from their land. We were to return to Reykjavík many times during our trip, but my first impression remained: a bustling place – but tempered with understanding, with humanity. A striking town, certainly, with its open streets of white-painted houses, the vast majority made of concrete, with corrugated iron roofs, for at the time of our visit Iceland had virtually no trees, and timber had – and still has – to be imported.

Perhaps the best view of the city is from the central lake: occupied at the time we arrived by a screaming, whirling horde of Arctic terns, all trying to find nesting space on its tiny islets: darting, stalling, diving – skimming across the backcloth of the city itself. Rising beyond the lake are the white and pastel houses with their red-lead roofs, clean and brisk-looking around the lake and harbour. Eider ducks busied themselves on the small lakes in the many parks across the city; dunlin and knot fossick below the harbour wall. The whole town feels clean and active. High above the city are the cooling reservoirs, for Iceland was at last beginning to exploit her rather off-beat natural resources by harnessing some of the geothermal energy. Hot water from some of the small geysers and hot springs are channelled to these reservoirs and used to provide hot water and central heating to all the houses and industries. (Further inland, at Hveragerði, the same natural heat was being used to power glasshouses so that some 300 miles south of the permanent ice of the Arctic, pineapples flourished, with bananas, peaches and other exotic fruits.) There was little to tell the visitor that near Reykjavík, at Kevlavík, was an American Air Force base with a population at that time at least half as big as that of the city itself. It did not obtrude: despite, or perhaps because of, the fact that the Icelandic economy relied at that time heavily on the American support, the place remained uniquely Icelandic.

Chris and Tony got back to the harbour before us – and they hardly looked happy. Before ever we left England, we had arranged transport for ourselves and our equipment straight out to our study site on the central plateau. We had hired a big, four-wheel-drive bus to haul us out along the tourist road to Gullfoss and on up into the central highlands. We were to be working in the Þjórsárver, a big water meadow some 20 miles across, tucked neatly beneath the Hofsjökull (High Glacier) on the central massif. The vast meadow of sphagnum and sedge, criss-crossed with streams springing from the glacier, is the main breeding site of the Icelandic pinkfoot – it was here, too, that earlier expeditions, such as that run by Sir Peter Scott and James Fisher in 1952, had come to work on the birds. But now it looked as if we might not make it. The phlegmatic Jonasson[2] was not prepared to take us in. The thaw, it seemed, had been late this year, and although the roads were now clear of snow the route was still in the grip of permafrost. That is to say that the gravel or moraines beneath the track were still frozen solid, but pressure from the weight of a vehicle would cause it to melt and mire the vehicle concerned. The most Jonasson could offer was to take us up part way, to another site, and then come out again later to collect us and take us on into the Þjórsárver. And it would cost us the double trip.

Seething with frustration, we had little choice but to agree. Time was short, for the Icelandic summer is a brief one and the snows would be in again by mid-September. So we accepted the inevitable and settled to the task of sorting through all our supplies to break them into two lots: sorting out sufficient food and essential equipment to last us the first few weeks before we could press on to the Þjórsárver itself.

A night in the excellent campsite in Reykjavík – and an early start. Our change of schedule meant that we would head for a little place called Fossrófulækur: a ford across a stream some 30 miles short of the Þjórsárver, and safe on the eastern side of the Kerlingarfjöll mountains.

Although the place is graced with a name, there is no permanent settlement; indeed, it is little more than a name on a map. A solitary

---

2 In Iceland, as in some other Scandinavian countries, surnames are not permanent within a family. The second name is always a patronym: thus Guðmundsson or Guðmundsdóttir – son or daughter of Guðmund … who himself may be Guðmund Finnurson.

hut stands on the stream bank and marks it as a posting point. A short way further down the stream, the water plunges into a steep and narrow gorge. Along the top of this gorge, on the basalt stacks which tower above the water, a small group of pinkfeet nest – outliers to the main population in the Þjórsárver. It would be worth a look and if nothing else, would serve for us to start to get the feel of things. At all events it would be better than kicking our heels in Reykjavík, despite the hospitable reputation of the Icelandic girls.

Jonasson drove the big four-wheel-drive bus himself. Perhaps feeling slightly sorry for us, he'd brought along one of his tourist 'guides' for the trip. A pity that neither of them spoke much English, but the trip itself was compensation enough. We rolled out of Reykjavík and off the tarmac. Past Geysir – perhaps the best-known of Iceland's attractions for tourists, yet in reality, something of an anticlimax: a few desultory steam spouts hissing away behind a barbed-wire fence. The Great Geysir, which used to erupt every so often with a jet of steam hundreds of feet high, was dead now: throttled with the tons of detergent poured into it over the years to make it oblige. Houses thinned; we had left the last major township behind at Hveragerði of the heated greenhouses. For the most part now, the bus wound over bare rock or hard-packed soil surrounded by desolate bog-meadows or deserts of dry lava. Redshanks and whimbrel called from the lonely landscape; black-tailed godwits, ringed plovers and ptarmigan scattered from in front of our wheels until we pulled off the road mid-morning at the head of the Gullfoss, Iceland's Golden Waterfall.

Gullfoss is justly reputed to be one of the most beautiful waterfalls in the world. The waters of the Hvítá, gathered together from a myriad of little glacial meltstreams from under the Hofsjökull glacier – from the very meadows in which we were to be working – plunge steeply through a narrow gorge, crashing 160 feet in a double span. The top 'flight' of the fall is impressive enough as one stands above it gazing nervously down, but its lower leap into a deep chasm is spectacular in the extreme – throwing rainbows into its spray which reaches up 100 feet or more. It is remarkable that one could then approach right up to it: to stand just above or right below. No written description can really do it justice: the noise, the power and the spray of it.

Jonasson had clearly been disturbed that we had come to Iceland merely to work. Satisfied now that he had shown us at least something of the island's stark beauty, he started the bus and prepared to move on. The tourist route stops at Gullfoss. Even the hard-packed soil which in those days made pretence of a road this far went no further. From here on across the central plateau to Varmahlíð and Akureyri in the north, the trail is marked clearly only on the map.

Jonasson pulled out his two-way radio and kept in constant touch with other vehicles along the route, checking on the weather and road conditions ahead. Every few miles we would lurch to a halt and, despite the language barrier, we soon cottoned on to the routine as we dug out the bogged wheels or threw heavy lava boulders into the ruts ahead so that the bus could crawl forward once more. These road-making stops became a regular feature of the next 60 miles, and we soon became used to piling out of the bus to reduce its weight as we teetered over flimsy suspension bridges. We didn't mind at all: we were revelling in the desolation and the birdlife.

We crawled steadily onwards; now even the bridges failed: the cost of throwing bridges across the many rivers had dictated to a country with little money that they were to be built only on routes with regular or heavy traffic. On roads in the interior, a span was bridged only for passing a very deep gorge or an impassable torrent. The remainder were forded. But even these were not fords as we knew them: there was no concrete base, and a vehicle had to pick its way across the shifting boulders of the natural stream bed. Further, the Icelanders' definition of a torrent is not the same as ours, and many of the 'fords' crossed raging waters so deep and so fast that we occasionally spied family saloons literally afloat in the race and only disgorged on the opposite bank some considerable distance downstream. Indeed this was apparently the norm, the expected; so much so that the roads on either side of such a stream are deliberately set askew from each other to allow for this same drift.

More birds: harlequin ducks now, golden plovers wheeling above the meadows, ptarmigan – Iceland's only gamebird – and merlins. Still, everywhere, whimbrel and the little red-necked phalaropes. And our first view ashore of a glacier: one of the outfalls of the Langjökull, brooding, distant and terribly grey, at the far end of Hvítavatn. The whole landscape, grey and desolate: a true desert of volcanic sputum, wild and incredibly vast. And then, as we straightened our backs after another road-mending stop, the sun caught the glistening top of a perfect sugarloaf icecap: the Hofsjökull and the end of our journey.

Although Iceland is of sufficiently high latitude that in the summer the sun never truly sets, there is nonetheless an extended twilight period as the sun dips towards the horizon and lifts again. In this red twilight the Hofsjökull looked almost unreal. The last dozen miles seemed to pass in a trance as we dropped down to the ford at Fossrófulækur. This Hofsjökull was to dominate our landscape and our lives for the next 13 weeks as we camped and worked in the meadows at its feet. No matter where we worked or moved, still it was there, watching over us, implacable. But somehow it was more impressive now, at a distance, than when we were close under its shadow or high up on its ice. After we'd pitched the tents, we stood and gazed at it in the dwindling light, poured a solemn libation of good malt whisky to its guardian spirits – and crawled to bed.

Two

FOSSRÓFULÆKUR: WHOOPER SWANS AND A FEW PINKFEET

I woke early the next morning and took advantage of a glimmer of sunshine to have a look around the little hollow that was to be our home for the next few weeks. Fossrófulækur was in fact one of the regular posting places for the annual sheep round-up. The central massif of Iceland is an area too rugged and inhospitable for conventional farming; instead, scores of sheep are turned out into the area as the snows retreat, to fend for themselves as best they may. Every autumn they are rounded up and taken off the plateau to be sorted and branded. But the sheep may range over literally hundreds of square miles, and with such a large area to cover these round-ups, or *réttir* as they are called, are protracted affairs, with weeks of tough riding over rough country chasing wild and wily old ewes. All the farmers cooperate in the round-up, and the ret becomes something of a social occasion – a hard-working, hard-drinking spree. At intervals, stopover points are set up for holding the sheep gathered to date, and the ford at Fossrófulækur was one such: a grassy hollow carved out by the stream, it was ideal for its purpose. The sheepmen had built two crude wooden huts here: the grander of the two, instantly recognisable by its smell, was for accommodating the sheep, the smaller a shelter for the herders. Like mountain bothies in Scotland, these shelters were never locked and offered a roof to anyone passing. Planning to use the herders' hut for cooking and storage, we had pitched our tents beside it on the springy turf. Looking around on this first morning, it really was an idyllic spot. The early sunshine caught the water of the little stream as it hurtled down to the pinkfoot gorge, and a harlequin duck bobbed in the current.

And that was another advantage of the spot: the stream was clear fresh water, something of a premium around here, where most of the streams are formed from the meltwater of snow or glacier – and glacier water has the reputation for doing nasty things to one's digestive system: a reputation we found later to be fully justified.

I hauled myself out of my sleeping bag and wandered away with the latrine spade up onto the moraine. This grey, gravelly material, characteristic of the glacial landscape, dominates much of the central massif. As the glaciers grip the volcanic rock, the underlying rock crumbles, and fragments are carried forward as the ice moves ever forwards. Then, as the ice melts or the glaciers recede, that till is deposited to form these moraine banks.

From a distance the moraines look grey and lifeless, but as I wandered across the gravelly slopes, picking my way between the larger boulders, the whole area seemed to be in flower. Wedged in the crevices, rooted deep in the seemingly sterile grit, was an abundance of little alpine plants – gentians, saxifrages and stitchworts – packed together in the grey chips. As is characteristic of all Arctic or montane vegetation, they were plants in miniature, but with a profusion of tiny flowers the like of which I had never seen before. There are times when I wish I were a better botanist, to appreciate and describe it all better.

By the time I had returned to the huts, the others were awake. Tony, appointed expedition cook (largely by default: no one else dared risk the abuse when things went wrong!) was starting the morning porridge. At this point, for those unfamiliar with expedition rations, I might perhaps be permitted a digression. The intention in provisioning an expedition is to provide the most nutritious, filling, yet balanced diet possible at a minimum of cost and a minimum of weight; remember, you have to carry everything with you. As a result, the cuisine, though doubtless extremely nourishing, is frankly appalling. One's appetite somehow seems to fail when confronted with the endless messes of porridge oats, Smash, Ryvita and Complan! The food is perhaps one of the greatest detractions of any expedition. I suspect the repetitiousness of the limited diet is a large part of the problem. Still Tony managed very inventively – and you can only die once.

Breakfast completed (and another tip for the busy expedition cook: for removing burnt porridge from the bottom of an old saucepan *nothing* works better than a piece of grit-embedded sphagnum moss), we set off downstream for the pinkfoot gorge. And downstream, quite literally, it became as we swung to and fro across the river to reach the foot of the gorge. This gorge restrains the lower reaches of the Blágnípa as it opens into the Jökulfall. The sheer rock walls are irregular and craggy and many tall stacks stand completely isolated from the cliffs themselves. On the tops of the cliffs and atop these stacks breeds a small colony of pinkfooted geese. This clifftop nesting habit is a strange one, for these little geese are usually meadow breeders. Perhaps for a smaller colony like this one cliff-nesting offers a greater degree of protection against predators, for such a colony would not be able to rely on the normal advance warning system of the larger breeding groups: wide views over uninterrupted terrain and the alarm calls of a thousand other geese. Whatever the reason the geese nest high on the stacks – and the goslings, once hatched, must plunge 80 feet down into the raging waters of the gorge.

We made our way up the bluff to stand at the top of the gorge itself, and prepared ourselves to count nests, eggs and chicks. Old-established nest sites are easy to recognise: the ring of droppings deposited around the nest by the incubating female causes a ring of darker, richer grass in the impoverished turf. But the droppings themselves persist; they do not decay as readily as in more temperate climes, and one cannot judge a nest as occupied just because of their presence. We moved steadily up the gorge, counting. But already we were too late – many nests were marked only by the scatter or broken eggshells. A couple of pairs of birds still sat tight, but to judge by the desertion of the rest of the gorge, these late-sitters were probably guarding addled eggs; most of the chicks had hatched and gone.

I suppose we might have been disheartened by such a setback, but it's difficult to remain dejected for long on a first field day in a new area. There was so much to see and record. Ravens in the gorge, snow buntings pottering like so many wagtails on the shingle banks of the stream itself and overhead, merlins and an Iceland falcon hunting over the short turf. I could cheerfully have stayed there all day, indeed all week, but we were there to work and over a quick cup of coffee from the flask we decided to strike out in the opposite direction to see if we could make contact with pinkfeet in some of the small meadows to the west of the Hofsjökull – just over the glacier, so to speak, from the Þjórsárver itself. Although not of the same scale as the central Þjórsárver – which protected some 3,000 breeding pairs then (and 10,000 pairs at the time of writing) – these smaller meadows might perhaps hold at least a few breeding pairs which we could approach.

The walk to Blágnýpuver, the meadow we had decided to head for, remains fixed in my mind, for in miniature it typified the kind of walking of which we were going to be doing a great deal in the following weeks, in terms of the types of terrain we were to cover and the general scenery. One of the things to regret about the whole expedition was how quickly one became familiar with even the most spectacular surroundings and so almost ceased to register them. But on this first trip all was still fresh and made a striking impression.

For most of the way we travelled along the tops of the moraine ridge, with firm and satisfying grip for the boots on the gritty surface. To the sides of us, the swift glacial streams whose periodic floods had added to these screes, and in the middle distance long ranges of rugged snow-capped mountains. Every so often we would need to dip down and wade buttock-deep across glacial rivers to cross to another moraine bluff. Everywhere, little pockets of snow and ice. Sometimes the path we took would follow a stream bed up through a vaulted cavern beneath the ice itself. These high, sculptured ice caves were incredibly beautiful: lofty, domed ceilings with tracery which many an English cathedral would be proud to boast; cool, peaceful – like limestone caverns but with real ice stalactites, light and shimmering. High waterfalls and rushing water; little patches of flowers in the spray zone; birds wheeling everywhere – birds so unaccustomed to human presence that they were ridiculously unafraid. And over all, the incredible calm and peace of it all – the empty moraines, the wild water and the distant, patient mountains.

As we dropped down into Blágnýpuver a pair of whooper swans moved off across the sedge, accompanied by their three young cygnets. Golden plover exploded from the tussocks, closely attended by little groups of dunlin. These small waders are often called the 'plover's page', and we could see how remarkably apt this was as they followed the plovers around. The colouring of the two species is remarkably similar, with stippled olive-and-gold back and the black head and bib of the breeding plumage. Perhaps the dunlin dance attendance on the larger plovers in reaction to some super-stimulus or exaggeration of their own kind?

Whoopers, plovers, dunlin and the little red-necked phalaropes spinning on the pools or chasing above the meadow in courtship flights. But no geese. We quartered the meadow and found another pair of whoopers on an unhatched clutch of three eggs, and eventually located a party of some 30 adult pinkfeet with young right at the far end of the meadow. But they were very shy, and clearly greatly disturbed by our presence; and by the same token, there was little we could usefully achieve in watching such a small sample of birds. It looked like we would just have to wait for our pinkfeet until we moved into the larger meadow of the Þjórsárver itself in three weeks' time.

We struck back over the moraine for home, for it was getting late: the dipping sun bathed the whole of the vast Kerlingarfjöll range in red light, just as it had caught the Hofsjökull ice the evening of our arrival. Somehow it seemed a long trudge back to Fossrófulækur. I don't think we were disappointed at finding so few geese; I think in honesty we were too tired to feel the disappointment as we tramped back over the yielding moraine screes towards the tents. A hot curry and coffee did much to restore morale as we sat around in the shepherds' hut for a council of war.

No pinkfeet, or at least very few; it was a bit of a blow, because not only would we lose some three weeks of our limited time in our enforced stopover at Fossrófulækur, but by the time we actually got into the Þjórsárver the young geese would be well-grown; we were losing not only time, but hard data. It couldn't be helped. Nonetheless we had to try and make the most productive use of the remaining time we were to stay at Fossrófulækur, and in the end we decided to split the expedition team into two. Tony, Adrian and Root would head south to explore some more distant meadows in the hopes of finding a more productive pinkfoot colony. And even if there were no pinkfeet around, there seemed to be plenty of whooper swans both in Blágnýpuver and in other meadows we had visited along the way; so to hedge our bets, Simon and I would mount a subsidiary study of the breeding behaviour of these swans, while Chris and Harry maintained a base camp.

Since we knew that at least one of the pairs of whooper swans in Blágnýpuver was still sitting, Simon and I elected to return to that meadow at once to study the parents' incubation behaviour. Once the eggs hatched, we could switch attention to some of the other families: the ages of the various broods seemed highly variable and we would be able to obtain information on the behaviour of families of different ages merely by moving from family to family – a more economical way of getting information than by following the development of a single brood.

The next morning therefore, we set off back over the moraine to Blágnýpuver, this time lugging heavy packs with stores, tent and sleeping bags, and a small hide. Apart from the first clear evening when we had arrived at Fossrófulækur, Iceland had not been kind to us with weather, and had honoured our presence with a persistent light rain for much of the time. Since we appeared undaunted she now decided to fling at us everything she had. The temperature dropped lower and lower, and a knife-like wind drove sleet into our faces as we breasted the last ridge and dropped down into the meadow. We pitched the tent just below the ridge and started to assemble our hide. The thermometer steadied at −4°C and the wind whipped the canvas fabric of the hide as we tried to sling it over its pole frame. And then, horror of horrors, we discovered that even on a sunny day with no wind we would never have been able to erect it: in sorting out our stores back in Oxford we'd come away with the canvas for one hide and the support poles for another, larger model. We improvised, stringing up the canvas inside the frame with twine, and as Simon retired to the tent to brew coffee, I was awarded first watch. The adult swans had moved away when we approached with the hide, and now I settled down to await their return.

We had positioned the hide some 60 yards from the nest: a large pile of dead moss about three feet high and three feet across the base. Whooper nests are often raised on a little mound and are set on the edge of a small lake or series of pools on which the male rests, feeding and raising his head from time to time to scan the meadow every two to three minutes. The nest cup itself measures about 18 inches across and is 9 inches deep. In this case a clutch of three eggs rested in the bowl on a few small feathers, although one got the impression that these were not an integral part of the construction, but rather the result of the frequent preening of the incubating female.

When Simon came to relieve me after my 90-minute watch, the parents had still not returned to the nest, but were patrolling nervously in the meadow just beyond, calling restlessly. For fear the eggs should chill in the cold air, we moved the hide back closer to the tent, to allow the birds to return. At a distance of perhaps 150 yards it made watching that much more difficult, but at least it reduced disturbance to the swans themselves. The weather had now settled to the absolutely appalling: driving snow and a steady temperature of −4°C. The 90-minute stints we did in the hide became more and more tortured, cramped in that tiny cube, unable to straighten up, let alone move about, our window on the world a narrow slit in the canvas offering a restricted view of only the one small tableau; boots, socks and trousers wringing wet from the glacial stream-crossings of the journey out. Cold weather I can stand – indeed I actively enjoy – as long as I can move about. But those soaking wet swan watches were a trial. It didn't help that the swans themselves were doing very little and that each watch afforded little of interest.

We struggled on around the clock and halfway round again – the 24 hours of daylight of the Icelandic summer permitted a continuous watch. Chris came out from Fossrófulækur with more supplies, and stayed on with us to ease the watching. We'd moved the hide again so that it now actually butted up against the tent, keeping both hide and tent considerably warmer. Three of us to watch meant three hours off for every one and a half hours on, and life became again almost bearable. But for one thing. For lack of fresh water, Simon and I had been forced to use the glacial melt for our drinking supplies. Remembering dire warnings about glacial water, we had been careful to boil it, but the real problem with such water is not the beasties it may harbour, but the fact that it is milky-white, laden with rock dust, a fine suspension of particles eroded from the rocks over which the glaciers creep. It's rock dust that makes bodies of glacier water appear such a startling blue in the open air. But inside one's body, those particles, coating or scouring away at the lining of the gut, have a decidedly deleterious effect. Not in queasiness, or indeed in any feeling of malaise – merely through producing an incredible degree of flatulence. With hide and tent close adpressed against the weather and little other ventilation, the atmosphere in both might have been warm at last – but was, to say the least, a trifle foetid. Chris, who had brought fresh water with him from Fossrófulækur, so hadn't succumbed to the glacial melt, was somewhat loud in his plaints about Simon's and my regrettably regular lapses.

And still we watched those swans. We were amassing a fair amount of information about them, albeit based for the moment on a single pair. It seemed that the female did all the incubation; if undisturbed she remained on the nest almost all the time, apparently asleep. She left the nest every three hours or so to feed, moving across to join the male. While the female sat, he remained nearby, feeding in or beside the little pools around the next. Every few minutes he would raise his head to scan the surrounding meadow for intruders. When the female left the nest to join him, the birds ranged much further afield; even so they never ventured further than about 30 yards from the nest itself. The female returned to the nest after an absence of 30 minutes or so; the male followed her back, feeding as he went, until stations were resumed for the next three hours of incubation. The whole series of observations suggested a regular three-hour rhythmicity of behaviour, with some 150–160 minutes on the nest and 20–30 minutes feeding away from it.

The snow had stopped and now a cold mist shrouded the meadow. Eventually we could no longer see the nest, so we decided to break camp and leave the swans in peace. We returned, draggle-tailed, to the Fossrófulækur hut, now emblazoned by Harry with a vivid ensign proclaiming 'Tony's Café'. For Tony, Adrian and Root had returned from their foray further south – still no pinkfeet, although plenty of whooper swan records for the pool. The medicinal whisky supplies got a severe shock that evening as we crowded into the little hut. I don't remember now how the rough and tumble started; but I do recall that it ended with Harry perched high on the roof of the hut, hurling down abuse and preserving-alcohol on the unrepentant heads of the rest of us.

In sharp contrast, the next day dawned bright and sunny, and was unanimously declared a holiday. A good decision as it turned out, for in all the 14 weeks we were to stay in Iceland, this was one of only three days when the sun actually broke through. We lazed in the sunshine, soaking up the unfamiliar warmth on our bare skins; in the clear air and high latitudes, the sun, when it does shine, is remarkably hot. A great deal of grime was sluiced away as we bathed mother-naked in the sparkling Fossró stream, and we stretched out again on the short grass, feeling rather at peace with the world. The harlequin who regarded our stretch of river as his own was not quite sure what to make of our intrusion, but now settled on the bank a few yards away from Root's feet. The snow buntings were busy picking around our freshly 'laundered' shirts, and the camp skua, a regular visitor now, engaged his time and talents in fishing the scraps of burnt porridge off the stream bed. The tents steamed cheerfully in the sun. It was good to be warm and naked on the turf, for the foul weather of the previous days had ensured that everything was waterlogged and nothing could be dried; indeed, this became a recurring problem of the entire expedition: how to get one's drenched clothes dry in the confined spaces of the tents. I had already recognised in myself a growing distaste for climbing out of a warm sleeping bag at crack of dawn to force reluctant limbs into cold and soaking wet clothes.

I stretched out and rolled over – and sat up with a start. I must have dozed off, for I had never heard them arrive. But grazing idly between the prostrate forms of my still sleeping companions was a regular army of ponies; 27 in all, to be exact, all fully tacked up, with 7 saddled as riding ponies and the other 20 fully laden as pack-horses. Visitors! I moved across to the door of Tony's Café and peered inside; seven solid Icelanders looked up and grinned a welcome. The other recumbent forms around the camp were being rudely shocked into life as the grazing ponies pushed wet muzzles into bare backs, and we prepared to introduce ourselves. Two of our visitors spoke fair English, so communication was not too strained. All were on holiday, just trekking across the central plateau as the whim took them.

Somehow, with the advent of visitors, the whole camp took on a festive air. We talked idly in the sun. We were introduced to the idiosyncrasies of the Icelandic pony: a stocky, rugged animal not unlike a Dartmoor or Exmoor pony, but with some remarkable extra gaits. One of these is the *tölt*, which can be performed at speeds ranging from the equivalent of a walk to a canter. In the tölt the legs move in the same sequence as a walk; first the hindfoot then the forefoot on one side of the body, followed by the hindfoot and forefoot on the other side. At a faster tölt the pony rolls from side to side like a camel, as both legs on one side leave the ground together – indeed, just one foot will be on the ground at some points. This extraordinary method of progression (which may need to be taught, even though the ponies have a natural tendency towards it) greatly eases movement over rough terrain, and also makes an easier motion for the rider, allowing long and comfortable spells in the saddle. The other extra gait is the *flugskeið* (flying pace): very fast over short distances.

We sat quietly in the sunshine, smoking in companionable silence, but all too soon our visitors left as they had come; they had only meant to stop to water their ponies. The air was cooling rapidly as we pulled on a few of our drier clothes and went back into the hut to prepare another curry. Peace descended on the camp again, but it was a camp refreshed as these Icelanders were, typical of their nationality, delightful people: easy, charming, open and intensely alive. That brief visit served to shatter some of the doldrums that inevitably descend on all expeditions from time to time, even under the best of conditions, and we felt somehow purged. On any expedition with just a handful of people living under fairly testing conditions, frictions are bound to develop. The expedition members necessarily live very much in each other's pockets. There are no fresh faces to see or fresh people to talk with, and even minor irritations can become magnified out of all proportion in the tense atmosphere – and, in this particular instance, conditions had been more than usually trying. The frustrations of not being able to get on with our work were mounting – first, our having been unable to get straight into the Þjórsárver at the outset, and now that we seemed to be too late for the geese outside those meadows – and now exacerbated by the miserable weather we had had to date: a fortnight of clammy dampness, mist and drizzle, driving rain and snow; day after day of living and working in sodden clothing, a fortnight of enforced dilettantism, unable to press on with the work we had really come for. And that inevitable bickering – which, though inevitable, can nonetheless destroy an expedition – was already not far below the surface. But the holiday, the fresh, vital faces, the glimpse of sun dispelled at least some of the gloom, and the devils retreated.

Life was still pretty good the next morning when I set off across the moraine for some nearby meadows at Granunes to continue my observations on whooper swans. Now that we had some information on the behaviour of the parents before the eggs hatched, I wanted to concentrate my efforts on the behaviour of families after hatching, as the little cygnets grew. And we know that there was at least one family, with three youngsters, in some little meadow pockets over below Kjalfell. Over the next few days I was to spend most of my time in these miniature meadows, observing and recording, and I quickly came to love the little valleys. The dark Kjalfell – a typical flat-topped volcanic giant, black and basalt – brooded over a series of very private meadows, each cupped within a little hollow of the glacial moraine. The meadows themselves were wet: essentially marshes of sphagnum and sedge with drainage streams flowing swiftly between steep turf banks. The whole area was littered with little alpine flowers and crowded with life: dunlin and golden plover called everywhere, purple sandpipers skimmed away over the sedge. The little meadows seemed somehow very secret and almost unreal in their completeness. The sensation of enchantment was intensified by a strange stillness, almost an eeriness, as if time stood still in these tiny valleys. Every now and then a little wind scurried across the grass: strange winds, low, rustling in the grass while the upper air was still calm; local winds, cutting a swathe only a few yards, rushing past a little way away. And then again the stillness and the plovers calling. I spent many days alone down in these meadows watching the swans, watching and waiting. And always it seemed as though I were intruding on a world not meant for us.

This first morning I scrambled down off the moraine into the first of the meadows and was pleased to see an alert male whooper fly over to his family and herd them away. I waded across one of the little rivulets which meshed the valley, and followed patiently. The whoopers made for a series of small pools on the other side of the meadow and soon settled on the water. This much we had noticed before: although tremendously shy on land, letting any of us approach no closer than some 200–300 yards before moving away, the whoopers seemed to feel much safer on water and, once settled, allowed us to approach much more closely while remaining undisturbed. I made myself comfortable on a little mound of dry sphagnum and waited. Now unperturbed, the swans floated idly on their pools or paddled across the mud at the water's edge, grazing the grasses and sedge. The cygnets hurried from one parent to the other, keeping close to the parents in tight parties.

An Arctic fox trotted along the moraine ridge beyond the pools as unconcerned about the swans as they were of him. Suddenly he caught sight of me and stopped. Clearly bemused, he sat down to study me in more detail. He really was a beautiful beast – indeed the finest I ever saw in Iceland: a big animal with a thick, silky coat of black and white and an enormous ruff upon his shoulders. He watched me carefully; since I appeared not to move, he got up and trotted over to have a closer look. He sat down again some 10 feet from me and puzzled. It was only the breathtaking excitement of having him so close to me that prevented my dissolving into helpless laughter, for his face held a ludicrous expression of sheer astonishment and bewilderment. This thing hadn't been there when he'd passed this way before, surely – and even now he had no idea of what it could be. It smelt pretty nasty to be sure – and didn't look too good either, come to that – but what *was* it? He edged closer, shuffling forward on his bottom. I dared not even lift my camera. Finally, satisfied that whatever it was it was going to stay there, he scratched himself and trotted away – but clearly still unsure, because he stopped for many a backward glance before disappearing once more over the ridge.

I breathed out, unaware until that moment that I had been holding my breath. It was incredible how unafraid most of the animals and birds of this central wilderness were: entirely unaccustomed to humans except perhaps for the occasional herder on ponyback, which was something rather different anyway, they were remarkably free of fear. And always something to watch: even as the fox trotted away over the ridge, the next divertissement was staged. It was almost as in a good ballet where, even while most attention is focused on the central *pas de deux*, elsewhere on the stage mini-dramas were being acted out among the *corps de ballet*. A female red-necked phalarope flipped over my head and settled on the water. She spun briefly, first clockwise and then anticlockwise, picking at insects on the surface of the pool with her slender bill; this pirouetting in the water is typical of these dainty clockwork mannequins as they sit high in the water, twirling first one way and then the other. The little female splashed and took off, swooping low over the meadow. Soon she was joined by another and another, until seven of the delicate little birds were whirling across the meadow and around my head in an excited courtship chase. Phalaropes are delightful birds – and not only because they are beautifully coloured. They form one of the few instances in the bird world where the roles of the sexes are reversed. It is the brightly-coloured female who courts the drabber male, the male who sits on the eggs and raises the young; the flighty female, once her eggs are laid, is off to court another male or, when the season draws to a close, to join others of her liberated sex in little hen parties on the meadows.

The courting party darted away and I followed them, to leave the graceful swans in peace and not disturb them overmuch with my attentions. Still the whole meadow quivered with new life. As I explored this and other meadows over the next few days in search of further swan families, I kept on coming across nests of dunlin, Arctic tern, golden plover. In one plover's nest, the last of three eggs hatched as I watched it, and the bedraggled little chick shrugged its way out to join its brothers and sisters, their yellow-and-green-spreckled plumage offering amazingly good camouflage against the sphagnum moss. The chicks would not remain long within the nest: fully-feathered at birth, with strong legs, within an hour or so the little balls of fluff would be scuttling between the sedges in pursuit of their parents.

Each little meadow, of some half a square mile, had its own whooper family and fortunately for me the cygnets were of a variety of ages from just hatched to a good few weeks old. As the days passed I formed a good picture of family life. In general, the range of the birds remains small, centred around a 'safe' stretch of water to which the family retires when disturbed. Around this base, the swans range out over the meadow, feeding on the sedges and other vegetation. The cygnets stay close to their parents, particularly their mother. The male may wander further from the rest of the family, grazing deeper into the meadow. But still it is he who acts as look-out, frequently raising his long neck to scan the area for intruders. At dusk, such as it is, the family retreats to the banks of their safe water, and while the cygnets are still young the female broods them through the two or three hours of the short Arctic night.

Although the adult pair will have been silent while incubating their eggs, they are now quite vocal, calling to each other across the meadow in strict antiphony. When resting on the water, the parents may duet to each other, each bird bowing its long neck, head to breast, and raising the head again with a mellow call. First one bows and lifts its head with a single clear note; its partner bows in turn and answers with a lower note, 'Whoo – per' 'Whoo – per', the clear call perfectly evoking the loneliness of these lovely meadows.

Whooper swans seem to be very successful in their breeding attempts: most broods consist of two, three or four cygnets, with an average brood size of three. Losses at this stage appear low – an impression confirmed by the reactions of the parents to potential predators such as marauding skuas, gyrfalcons and foxes. The family may group together and make for the safe pool at the heart of their range, but they are unhurried and watchful rather than alarmed. When I was there, only once was a male sufficiently perturbed by my presence to leave his family and fly low over the meadow to dive, threatening, down at me – and this was little more than a token gesture. But losses are greater when the birds leave the central highlands of Iceland and fly south to the coast before migrating to Scotland and northern England for the winter. Ian Newton, who collected data over a number of years on whooper families as they arrived in Scotland after migration, recorded an average brood size in that year of only two cygnets per pair – a loss of a third over our own pre-migration figures for that same year.

I really enjoyed those days in Granunes with the whoopers – but at last the time had come when we were to move on and make the second lap of our disrupted journey into the Þjórsárver. I breasted the high moraine ridge for the last time and headed back to the ford across the Fossro. Below me, away in the meadow, a pair of swans called to each other in the mist: 'Whoo – per' … 'Whoo- per.'

Three

ASGARD AND AN
UNEXPECTED JOURNEY

The long hiatus was over; in the habitual drizzle of the early morning, we stood by the roadside surrounded by our stack of equipment awaiting the return of our transport. And at last, to our considerable relief, Jonasson's large four-wheel-drive bus was amongst us, ready loaded with all the extra stores we had left behind, all prepared to take us on into the Þjórsárver. We had looked forward to this moment for so long – and soon enough we were off, on up the road to Asgard at the very foot of the Kerlingarfjöll mountains. The road wound up over the moraine until finally the great bus lumbered through the swollen waters of the Jökulfall, cascading brown and turbulent through its narrow gorge, to reach the end of the marked road. From here it was strictly cross-country without even the pretence of a trail on the map; the bus must pick its way like a sensitive pony through the jumbled lava, bogland and moraine of the glacial plain. Jonasson swung the wheel and we began to lurch across the gravelly moraine.

We'd gone perhaps 20 yards before we stopped, axle-deep in the waterlogged soil. Jonasson was firm: he was not going to take his bus any further. Indeed, to make his point, he was already beginning to unload all our gear from the back of the vehicle. Reluctantly we helped him pile our crates and cases on the roadside – and then he was gone; it seemed he couldn't be gone quick enough. And we: we were left in a dejected circle wondering what to do now. For this we had waited three weeks, yet were still 20 miles short of our objective, 20 miles of rough and inhospitable country – and we now no longer even had the cold comfort of imagining we would be going into the Þjórsárver a few weeks later. A drink seemed to be called for, and we made our way up to the little ski hostel huddled against the mountainside at the end of the road. Skiing is a popular pursuit for Icelanders and visitors alike, and despite its isolation the little ski hut was warm and welcoming. None of us realised that on Asgard's slopes we were on a still-active volcano that was to erupt a few brief weeks later.

We settled into a corner, nursing our drinks to try and work some way out of the mess. This expedition wasn't going quite the way we had planned it. As the mid-July snow lashed against the windows we reviewed the alternatives. There was no way we were going to walk in – not over that terrain with over a ton of stores and equipment on our backs. But what was left? We knew that there was an expedition due sometime towards the end of the month from Radley College – indeed they had already agreed to link up with us out in the Þjórsárver to provide the extra manpower we would need to catch the geese for blood-sampling. They would surely help us in? But if we couldn't get in there was no guarantee that they would either; anyway, we had no idea when they were due and we couldn't just sit about kicking our heels for another two or three weeks waiting for them to turn up. We'd wasted enough time; we wanted to go in now. Perhaps we could return to Reykjavík and try to get back up to the meadow on the other side of the Þjórsá river, from the small town of Hella? But the roads were likely to be no better, and we'd still be left with a very long walk. Perhaps, then, we could hire some ponies and trek in that way? The ski hut proprietor shook his head: there weren't any ponies on this side of the river; we'd have to go back south to collect some, and it would take a long time to walk them back. Besides, to carry all our gear, we'd need a whole string of ponies – and each ponyman would need an extra two animals to carry his own food and tent. The prospect looked more and more grim. What other choices did we have?

For three days it snowed and sleeted – the wet metal of our mess tins froze to our hands when we tried to wash them in the cold water of the streams. For want of any better way of using our time we fell back into the routines we'd adopted long before at Fossrófulækur. Every day we would split up and scour the surrounding meadows for whoopers and pinkfeet. We got colder and colder, and more and more miserable. We'd taken over an empty Nissen hut to store our abandoned kit, using a small corner of it for our own daytime use – but our clothes, permanently soaked, began literally to rot on our backs. Everything smelt of mildew and decay. Morale was at a pretty low ebb. Already we had been at Asgard for a whole week and we were getting nowhere; the forced inactivity began to breed its little irritations. All of us became sharp-tempered and short of patience; there came occasions when I had to walk away from the camp to save myself from making a bitter riposte to some imagined slight. It was becoming ridiculous: the expedition was beginning to fall apart in tune with our clothing. Finally, I made a decision: we would retain a base camp here at Asgard, and Adrian, Root and I would try to walk in to the Þjórsárver. If we could make it unladen, then we could all move in, with the bare minimum of stores and equipment; we could then at least start the behavioural part of the pinkfoot work – waiting for the support of the Radley expedition and their Land Rovers to bring in the rest of the gear when we started catching.

I left the Nissen hut and went back down the valley to my tent before I had time to regret the decision. It was going to be a long walk in: some 20 miles each way. And going unladen, we'd have to do the complete round trip in a single day. Still, we had no other real options; I drifted into an uneasy sleep. Typically, I woke the next morning with a bad attack of diarrhoea, but I plugged it with Enterovioform and set off with Root and Adrian as planned.

We moved away up the Jökulfall and struck out across the alluvial plain at the foot of Loðmundur, one of the peaks of the Kerlingarfjöll range. After an initial scramble over the rugged jumble of lava boulders just behind our camp, the going was fairly easy: a steady swing across the meadowlands under the mountains. We splashed through the two or three small streams which quartered the meadow, and trudged on. Purple sandpipers flicked away from us and skimmed over the grass. Clearly many of them still had nests, for they put on a superb performance of distraction displays a few yards ahead of us as we walked, feigning injury and calling to draw us away from their eggs and young. A bird would flutter ahead of us in attempted flight and collapse on the ground, calling plaintively. As we approached she would stumble away, trailing one wing at an unbelievable angle behind her, to draw us on. Sometimes, too, a bird would fold both wings tight across its back and scuttle through the grass a few yards in front of us, so like the movement of a rat or other small mammal that it was difficult to believe it only had two legs. As we walked past, ignoring them totally, they would fly forward beyond us again – a perfectly normal flight – to collapse once more in a fluster of 'broken' wings on the grass – an amazing display to distract predators.

Our attention drawn to the ground, we noticed too a strange formation, often repeated, of concentric rings of moss, each separated from the next by the bare black silt of the plain – a formation apparently characteristic of that meadow, for we never saw it anywhere else, before or after. Until we found one with drifts of snow piled up between the moss ridges, we hadn't been able even to hazard a guess as to how they might have arisen. Now we still didn't know, but could at least offer a plausible explanation: we supposed that the ring formation must arise in much the same way as a mushroom ring – the mosses of the central clump cast their spores outwards; these would tend to land in a ring around the central hub. When these plants in their turn tossed their spores into the wind, these would be equally distributed back towards the central core and out again into another outer ring and thus a series of concentric rings would begin to grow. While some spores would doubtless fall between the rings, we had already seen that the major ridges trapped drifts of snow between them. These might well kill off the few between-ring intruders, leaving the bare earth which characterised the formation. This might well not be the true cause, though, for why had we not observed such rings elsewhere? But, satisfied that we could offer at least some plausible hypothesis, we plodded on.

Up over a high buttress of the towering Kerlingarfjöll, we crossed a deep gully on a natural bridge formed of compacted snow left behind after the melt. The single span, eroded from below by meltwater, arched gracefully across the gully; below us the fast-flowing meltstream hurtled down the mountainside. Up and round again – and out onto a bare plain stretching mile upon mile ahead.

Two hours up. We stopped for a smoke and a drink from one of the many streams, contemplating the desert ahead. Pure lava desert – and while the apparent lifelessness of the grey moraines had been given the lie by their myriad of little alpine plants, this desert was dead indeed: a waste of razor-sharp lava stretching away into the distance. Picking your way across this stuff is like walking across acres of clinker scraped from the bowels of some diabolical coke furnace. The sharp, mis-shapen masses of welded tuff turn your ankles mercilessly as you stumble forward. Two more endless hours before we were clear of this lifeless ash and climbing down again onto the moraine of a wide glacial plain.

Over to our left, a great tongue of ice stretched out over the plain from the Hofsjökull glacier. The braidings of its meltstreams formed countless little rivulets across the moraine. A score of times and more we dipped down into these little streams to wade across. By then we were so wet anyway that we were almost beyond caring, and besides, none was more than crutch deep. This Hnifá plain seemed almost as empty of life as the lava desert we had just crossed – as if we had left the real world back on the plains below Loðmundur. We trudged on, rarely talking now, saving our energies for the walk –for 'plain' is perhaps something of a misnomer: the ridges and valleys in the moraine were often as much as 20 or 30 feet high from crest to trough. It might be 20 miles into the Þjórsárver from Asgard as the crow flies, but it was something very different on the ground.

The journey seemed destined to break into two-hour hauls, for it was six hours after we had started that we waded through the final deep stream and breasted the last rise to look down onto the Þjórsárver. As if in token that at last we understood what Iceland required of us, a party of pinkfeet moved slowly away from us across the moraine. The vast meadowlands lay beneath us: mile upon mile of green sedge and wetland. As we stood there, drinking it in, there flooded into our separate minds the lines from J.R.R. Tolkien's *Hobbit*, when Bilbo Baggins first stands above Esgaroth:

The lands opened wide about him, filled with the waters of the river, which broke up and wandered in a hundred winding courses, or halted in marshes and pools dotted with isles on every side; but still a strong water flowed on steadily through the midst.

This was indeed what we had come for. We turned away in awestruck silence, to return the way we had come.

The long walk back tested every tendon and every shred of willpower in our bodies. We really pushed the pace, but we were already very tired. Breaks were limited to five minutes in every hour and we stumbled on, staggering the steps of blind exhaustion. Back across that hellish lava, down onto the Loðmundur plain and on. And then, after 12 hours of solid walking from when we had set out, we were back at the camp, dropping onto the soft turf and listening to Chris, with all the charm of a grateful diplomat, berating us soundly and abusively. What we had done, or hadn't done, I never learnt. Still, we'd made it. The route was passable and we could walk in – even, we reckoned, carrying light packs. It was going to take some working out, though. A tent and food for a day for two people weighed about 40 pounds, and that already makes up a full pack for that type of country.[3] Clearly we could not carry stores and equipment in with a stopover in the middle, for each trip would then only provide perhaps some 60 pounds of equipment between the six of us.

We would have no choice but to make the 40-mile round trip each time, for by leaving the tents at Kerlingarfjöll we could then take two or three times as much kit on the drop-off runs. Eventually we compromised: Chris and Simon would move out to the meadow with one tent, to start behavioural observations. The rest of us would operate a shuttle service, ferrying in equipment and supplies until we had accumulated enough for us all to stay.

In the end, we really got quite used to those walks – so much so that in the memory they all seem to fuse together into one. The route, though unmarked, became depressingly familiar: we quickly became accustomed to the countless immersions in milky, silt-laden meltstreams which crossed our path (a fact to which our underwear

---

3 These were not the lightweight tents of modern fabrics which have now been developed; ours were heavy canvas tents designed specifically for the Arctic.

bore mute testimony for many months afterwards), to the mindless trudging across the soft gravel of the moraines, boots sinking deep at every step and tugging mercilessly at one's heels as one wrenched them out again, the ankle-wrenching confusion of the lava. Indeed, the most impressive thing about that whole period was the sheer boredom of the walking: hour upon hour of rhythmic trudging across a landscape that barely changed, mind deliberately emptied, almost as a defence against the discomfort and boredom. Just walking.

It is incredible the things with which you occupy your mind on a routine march like that: despite general belief, one's head is not filled with clever scientific thoughts, or deep philosophical musings. Rather, one dwells upon the delicate texture and moist taste of a Marks & Spencer's soft bap tenderly filled with scrambled egg, or the splendid and sensual feel of freshly-kneaded bread dough. It is remarkable what you find you *do* miss on an expedition – and how often you think about food!

The mindlessness of those walks, an inevitable consequence of the sheer mechanical boredom of it all, was nonetheless a real danger. You can take nothing for granted in country like that, and there are tremendous dangers in becoming too blasé or being only half-conscious. We became, as I say, distressingly familiar with the routine of crossing the many fast glacial streams. We started off using ropes to cross all but the smallest, but we didn't really need them and the precaution soon lapsed. We could manage just as well by angling across the current in a chevron pattern upstream, probing the loose and boulder-moving stream bed with a short pole; and anyway, the ropes were heavy and an unwelcome added burden. So we stopped bothering. But Iceland is a country of sudden change: a shallow stream, calf-deep at the most when you crossed it on the way out, could, in three hours of hazy sunshine and melt, become a raging torrent, breast-high, by the time you returned. We were caught out that way more than once. We went back to using ropes for the larger rivers.

Although the many trips blend together, it's that sort of occasion that sticks in your mind: slow-motion ciné shots of sudden accidents – the alarm, and the people moving in such incredible slow motion in their efforts to assist. The time that Christopher, back with us for one supply trip, swept away from us down a turbulent Hnifá as we crossed chest-deep in the troubled stream. For once, we were even roped, but he lost his footing on the shifting boulders of the bottom and was plucked from the rope like a marionette. Fortunately the Hnifá follows a very tortured course across the plain, and we picked him up some 150 yards downstream, becalmed in the slack water of a bend, winded – but cursing. Or the time when the soft gravel of a pan resolved itself as a 3-inch skin below Root's heavy boots and plunged him knee-deep into the boiling mud pool it encrusted. Or again: I can feel now the impotence, the heart-stopping helplessness of sheer exhaustion as, at the end of yet another 40-mile round trip, particularly tired after battling against a Force 9 gale most of the way, we came round the final ridge of the mountain – no more than 200 yards from our base camp at Asgard – to see the canvas of the big framed store tent flap once as if trying out its wings and then lift bodily into the teeth of the wind and disappear over the horizon. It *had* to wait, of course, for that precise moment, so that we were back in time to observe it but impotent to do anything about it. (Such incidents, it must be added, did little to gladden the heart of our insurance agent when we returned home.)

But all this, and even these long supply treks into the Þjórsárver were still to come. For now, we had proved that we could do it, and we were in the mood to rejoice. We declared a two-day holiday and the need for a bath – as the last we might get for a number of months – and set off for the hot springs of nearby Hveravellir.

Over most of Iceland, hot spring areas are essentially vast lakes of boiling mud through which hissing steam spouts force their way to freedom. The slow, rolling boil of the mud pools, the gouts flung high by the escaping steam, resemble nothing so much as a boiling chocolate custard. The air is thick with steam from the mud itself and a heavy, sulphurous shroud hangs over the area. Hveravellir is different. It has its share of boiling mud vats, to be sure, but much of the area has a firm surface of white, quartz-like material. As a result, many of the steam spouts are not confined within the mud pools themselves, but hiss up through vents in the solid ground. Over the years, material has condensed out of these hissing pillars of steam to form a smooth and solid cone around the vent, like a hollow stalagmite. These cones are of the same bright white material as the ground itself, the mouth of each vent dusted internally with yellow sulphur. From the smooth, polished bore, great jets of steam thrust high into the air. As before, the whole area is shrouded in a heavy miasma of sulphurous steam, but at Hveravellir the individual spouts are clearly distinguishable in the general haze. The continuous roar is almost deafening, the glistening white landscape, dotted with bright, sulphur-topped cones, a breathtaking spectacle.

The area is sharply defined, as are many of the volcanic faults on the island. The pall of steam which hangs so heavy over Hveravellir has distinct limits, giving the spot an unreal appearance as you approach. The area is possibly much more developed now, but in those days a small rest house had been built on the edge of the springs for the convenience of visitors. There were a number of these *sæluhus* up and down the country, and they were quite characteristic buildings: airy, wooden structures with deep red corrugated iron roofs: the polished wood interiors of the two rooms maintained scrupulously clean and tremendously comfortable. Very attractive buildings, radiating their own welcome; we opened the door and settled comfortably into the unaccustomed luxury. Outside the window, the big steam cones hissed their continuous roar of vapour. Unlike geysers, which give brief, if spectacular, bursts of activity, these steam spouts are constant and continuous: a pillar of steam stands permanently above each vent, always alive and always in motion with an incredibly *real* power. The continuous noise, the great hissing jets of steam, combine to emphasise the awe-ful sensation of energy: a feeling of power suppressed, power contained, mingled with a feeling of suspicion that it is uneasily and only barely contained, that the whole might explode in a release of that power at any moment.

We wandered, awed ourselves, between the spouts. That's one of the most striking things about Iceland, especially back then, when there were few health and safety restrictions! You feel very small, very close to the ancient elements of earth, fire, air and water. Still, we had come for a cleansing of the body as much as a cleansing of the spirit: the boiling water gushing out of the ground cooled as it flowed over the bare rocks till it collected, at about blood heat, in a natural basin in the rock, whose walls had been built up with concrete and stone to form a deeper bathtub.

Here we lazed, deep in the steam, soaking out of our jaded pores the tiredness and grime of the last few weeks: trading it for a lingering taint of sulphur! This natural steam bath lifted the exhaustion and frustration from us; the next day should at last mark the true start of the expedition for us.

And so those long supply treks began, until eventually, some ten days later, the last of us pitched our tents on the green hillside of Nautalda in the Þjórsárver.

Four

HEIÐAGÆS

We had a somewhat restless and damp night of it, and early next morning moved the tent from where we had unknowingly pitched it – on top of the underground springs which give this side of the Þjórsárver its only fresh water. Still, at least we had now found the springs whose presence was the very reason we had elected to make Nautalda our base; we'd been looking for them all the previous week! Having re-pitched the tent, we sat outside on the short turf for breakfast. Nautalda is a little T-shaped spur rising up out of the Hnifá plain, the short headpiece of the T fronting onto the very edge of the Þjórsárver meadow. Below us, as we attacked the interminable porridge, stretched the open meadowlands: 20 square miles of braided waterways, of sphagnum bog and sedges – and 30,000 geese. All along the left side of us as we sat ran the moraine and ice of the Hofsjökull; ahead, the first ridges of the great Vatnajökull, 30-odd miles away. And, finally, to the right, the waters gathered together to form the Þjórsá in the distance over the meadow. A vast, treeless wasteland, desolate, mercifully desolate: no human settlement for at least 20 or 30 miles in any direction. Silent but for the moving waters, the creaking of the great icecap and the plaintive cries of whimbrel and golden plover. And all around us the pinkfeet moved steadily and purposefully across the meadows, grazing, bathing or simply loafing.

Now at last this little goose should be properly introduced. The heiðagæs, the pinkfooted goose of Iceland and Spitsbergen, had long been considered to be merely a subspecies of the bean goose of western Europe: a smaller bird – indeed, the smallest of the grey geese – with a short, flesh-pink bill and distinctly pink legs and feet. However, more recently the pinkfoot was recognised, at least by most biologists, as a separate species. 'Grey goose' is something of a misnomer, for the pinkfoot is an extraordinarily handsome little bird with a barred plumage of chocolate and dove-grey, each feather tipped with white. It is more delicate, too, than the other grey geese: while they are heavy-bodied, rather squat-necked birds, the pinkfoot is far lighter in build, with a slim neck and shapely head. Its little bill is not the heavy wedge of these other birds, but more a little duck-like snub; it is this last feature which gives the pinkfoot its scientific name, *Anser brachyrhychus* (or short-billed goose). Even in its behaviour the pinkfoot is a delightfully elegant little bird as it picks its way lightly among the pools and streams of the meadows. Despite my scientific cynicism and initial low regard for geese in general, I found myself becoming remarkably fond of the little birds in the weeks that followed.

As we have noted before, we had come to Iceland because we suspected that not only was the pinkfoot a species distinct from the larger bean goose, but that the Icelandic race might itself prove distinct again from other geographical forms. As indicated earlier, the Icelandic pinkfoot breeds here in the Þjórsárver and winters in the north of Scotland, and another population breeds in Spitsbergen and the north of Norway, to winter in Denmark and northern Germany. Ringing studies have suggested that these two populations are geographically distinct: at no time do they have the opportunity to meet, or appear to intermingle. It was quite possible that the Icelandic pinkfoot was something unique – and if so, here was an exciting opportunity to observe evolution in action.

Ninety per cent or more of the breeding population in Iceland is centred in the Þjórsárver, and now, in 1969, plans had been put forward for a major hydroelectric scheme on the Þjórsá, which would flood the meadows and possibly wipe out the geese – for there was really nowhere else which could accommodate such numbers of breeding birds. Even before a more global push for greater reliance on renewable energy sources, there was no doubt at all that a relatively poor country like the Iceland of the time desperately needed to exploit all its resources of power and raw material. Despite the overwhelming impression of elemental power all over the island in steam spouts, geysers and volcanoes, it had at that time proved largely impractical to harness this geothermal energy: the geysers and spouts were unpredictable and temperamental; most of them were too powerful to harness at all with the technology of the time. Hydroelectricity was the obvious answer.

It seems odd in a 21st-century context to understand the limited options of the time: while we now have technologies for harnessing geothermal energy direct, they were far less sophisticated at that time. In the same way, while we now tend to harness water power through a series of efficient micro-hydro schemes, in those distant days, generation of hydroelectricity meant the engineering of major dams to hold back millions of tons of water. A scheme on the Þjórsá would mean flooding the entire area. But it wasn't as if it would involve flooding land useful to humans: all the great rivers that force their way down to the coast of Iceland have their origins deep in the interior of the island. All of this was, and is, uninhabited; most of it uninhabitable: vast wetland areas or tracts of ice and lava. We have already seen how the only truly hospitable part of this strange country is the narrow coastal strip of relatively more fertile alluvial silt. It was clear that nothing lived in the interior except the wild geese.

It is, as I say, difficult to put yourself back into the mindset of the time; society did not have the same awareness or concern for conservation as we do today, and given the limitations of 1960s technology, my sympathies were all with the Icelanders. I was certainly not going to let the geese stand in their way. But for one thing. It was not true that nothing lived in the interior except the wild geese. The Þjórsárver was a vast wilderness teeming with life: a great natural reserve preserved through its very inaccessibility. The area supports a staggering diversity of species of animals and plants, and so very few people had penetrated these areas at that time that most of these wild animals had very little fear of man and were ridiculously tame. So: flood parts of Iceland's interior for hydroelectric schemes by all means – but at least, in the European Year of Conservation, leave the Þjórsárver alone. If you like, we were using the pinkfeet and the possibility that they were a distinct subspecies as a front: to give specific scientific reason for not submerging these hauntingly beautiful meadows. The old argument of 'Oh, it's such a lovely area' holds little weight against the economic demands of power, but if we could demonstrate that the Þjórsárver held the only breeding population of a distinct and unique taxon, we might have a case which the politicians might listen to. So we had come to study the Icelandic brand of pinkfoot, to review its general behaviour and its breeding behaviour in these meadowlands – and later to take blood samples back for analysis to try and establish its true position as distinct from – or perhaps identical to – the other pinkfeet of the world.

The previous evening we had sat late into the long Arctic twilight, discussing how best to set about our task. Eventually we had decided to split ourselves into three separate parties, each of which would cover a different area of the meadow. So this morning, with breakfast over, Root and I helped the others break camp and load packs, and saw them off across the meadow in the thick misty drizzle before we returned to our tent to start our own watches here at Nautalda.

We soon slipped into our watching routine: every 15 minutes during the day we scanned the meadow with binoculars to the limit of our vision, counting the numbers of geese and the numbers involved in each of a series of recognisable behaviours: feeding, drinking, walking, alert, resting, preening, bathing and so on. This would give a picture of the general pattern of activity of all the geese across the whole day. Between the 15-minute spot surveys, we would make detailed observations of the behaviour of a small group of perhaps 20 geese based near the tent. These geese were close enough to observe much finer details of behaviour. Since Root and I were also charged with recording the behaviour we observed on film (both for illustrative purposes and also for frame-by-frame analysis of the behaviours shown when we were back in the UK) we had rigged up an annexe to the tent proper under a tarpaulin (we had long ago abandoned the hide). Under this tarpaulin we stretched for hours, recording and filming on our big 16mm camera.

Days passed – days when we could not leave the tent for fear of disturbing the geese we were there to watch – so we dined on

Enterovioform. We were working three-hour shifts: three hours on, three hours off, three hours on and off and on. The best times for filming were 4 am and 4 pm – and when one of us was filming the other had to make all the behavioural recordings. So we decided to shift to a 12-hour day: 3 hours watching, 3 hours filming, 3 hours filming, 3 hours asleep. With no light cues because of the continuous daylight, this adjustment proved much easier than anticipated; to our surprise our bodies adapted remarkably easily to the regime. For some time now we had been taking medical records: body temperature and other physiological changes. Now, there was no doubt about it: the flux of body temperature up and down over the normal day was condensing into a 12-hour period. Other measures agreed: we were well and truly on a 12-hour cycle.

Still the days went past; the pot-hooked notebook pages piled up under the tarpaulin. And still it rained: we had not had a day free of rain since that heatwave at Fossrófulækur. Our clothes were falling off us: soaking wet for months, they were now disintegrating rapidly. The tent filled with the musty, foetid smell of decay. Our increasing hatred of climbing out of a warm sleeping bag into wet and mouldy clothing finally got the better of us, and we dispensed with clothes. Indeed, in the succeeding weeks, even when we were outside the tents again and over the meadows herding the geese to capture points, we rarely resorted to more than a pair of boots and a donkey jacket. We must have looked very odd: naked, hairy savages (we were all in dire need of haircuts and a shave) prancing about over freezing bog-meadows; fortunately there was no one likely to see us.

Those weeks of watching were perhaps the most relaxed of the whole trip. Root and I got on very well with each other at any time (in fairness Root had a tremendous facility in getting on with *any*one), and now at last we were doing the work for which we had come. And still we waited and observed; and still it rained. When we eventually joined up with the others and looked back on those weeks, comparing notes, it was amazing how many experiences we had, unwittingly, shared. Like the unbelievable craving we all, independently, developed for fat: scooping great gobbets of margarine out of our 5lb tins with a spoon, or at best with a Rich Tea biscuit. We ate pounds of the stuff – I suppose it was in response to the cold, for we all put on a thick blubber-like layer of subcutaneous fat. (For my part I put on a full 35 pounds while I was in Iceland, which dropped off me with no effort on my part within a fortnight of returning home.) Or like the strange routines one falls into and the very real distress caused if one is forced to break them. The reading and re-reading of favourite passages from books. Root and I had retained two books only: Tolkien's *The Hobbit* (of course; this was 1969!) and Darwin's *Voyage of the Beagle*. Both of us had read them many times before we finally deserted them for the inspiring prose of the cooking instructions on our various freeze-dried meals.

As the weeks passed we were collecting a great deal of information on our geese. All the goslings had hatched, of course, before we had managed to reach the Þjórsárver, and now the pinkfeet moved across the meadow in little family groups, dawdling their way across the plateau. We found the geese to be night-active, with most concentrating on feeding during the darker hours and the majority sleeping around midday. Sleep periods were short, rarely lasting more than a couple of hours; as the geese settled to sleep or roused themselves, there was a tremendous flurry of preening and bathing all across the meadow. The families appeared not to have distinct, fixed home areas but moved across the meadow in a huge circle; they maintain around them a movable 'territory' or personal space as they travel. The family groups were tightly knit: any gosling straying into the territory of another family will be driven off at once. Often, if an older gosling has been gone too long from its own family, it may have difficulty in being re-accepted. As a result, little creches form of these older goslings, whether orphaned or rejected in this manner.

Interestingly enough the behaviour of these 'orphan' goslings differs markedly from that of those remaining in family groups. One little creche attached itself to our campsite and haunted the tent, so we had ample opportunity to watch them. It was clear that these little birds chose a fixed home territory rather than moving around the meadow like the family groups, and that – again unlike those family groups – they were day-active: at night, they piled together for warmth and slept in great heaps of up to 40 birds.

It wasn't just the geese that frequented our camp. One of the compensations for the days spent cooped up in that foetid tent was the incredible variety of the visitors and camp-followers. We had regular friends: the wheatear that popped under the tarpaulin about mid-morning each day and pottered unconcernedly between the jumble of used film cans, flaunting his clean breeding plumage; the white wagtail which roosted each evening inside one of our billycans; the arctic skuas who became almost too bold in their piratical raids upon our stores. And still we stayed within our tent for fear of disturbing overmuch the little geese. Root and I were getting cramped by now – and sick of the sight of Enterovioform. We had read and reread the instructions on the Smash packets, and had now progressed to the dried runner bean packages. Enough was enough.

That last evening, the rain stopped. We supped on curried eggs – or at least a hot mixture of dried egg, curry powder and water – and then, on the moment, decided to break out: to wash up in the little stream below the camp and nurse a cup of coffee and a smoke out in the open air of the long, clear evening. We struggled out of the tent, bearing the oblong metal mess tins which every expedition camper regards as the only form of platter – and stopped in alarm. Number Five in my little catalogue of household hints to the expedition housewife (whatever their sex!): don't make curry from egg powder. Under the dim orange light within the Arctic Guinea tent, Root and I had tucked into this gourmet confection with enthusiasm; out here, in the true light of day, the scrapings in our mess tins were bright, bilious green. Green as the yellow-green sphagnum moss around us – unmistakably, horribly green.

Weakly, we tottered down to the stream and scrubbed away the evidence. Delight of delights, we cleaned our teeth for the first time in weeks. And then we sat down on the sloping banks of the stream to drink our coffee in the red dusk. It was a perfect evening: still, cool and full of peace. The meadow brimmed with life before us. And as we sat, a little family of pinkfeet moved slowly across the turf in front of us – closer than any we had seen from the tent: arrogantly, *totally* undisturbed by our presence. We lobbed a chunk of lime-green sphagnum at them in disgust, and retired to the tent.

Five

TO CATCH A GOOSE

Chris and Simon had arrived back; it was time to move on. An army of willing helpers were by this time camping back at Kerlingarfjöll, patiently waiting for the order to move in. For the next phase of our programme involved catching some pinkfeet, both for routine anatomical measurements and also so that we might take blood samples for later analysis. And to catch the geese we would need help.

David Hardy, biology master at Radley College, had offered that help. As well as being head of biology, David was also commandant of the college's ATC (army training centre) and regularly brought his student cadets to Iceland on exercises. Generously, he had put the services of himself and his charges, and a week of their time in Iceland, at our disposal. David indeed knew the Þjórsárver better than any of us, and it was largely at his instigation that we had set up our expedition in the first place. We'd accepted his offer of assistance with alacrity months ago, and now we were ready to move on it. The first thing we had to do was bring them in, and try to use the opportunity to carry in all the additional equipment we would need for the catch. So when the others had all returned to Nautalda, we packed up and headed back to the ski-hut at Asgard.

High up on the side of Nautalda, the hillside dips into a narrow dingle which slopes gently up towards the crown of the hill itself. The little gully forms a natural funnel, collecting together the undulations in the rise of the hill to channel them more easily up to the crest. At its tip it narrows, funnelling still tighter and within this funnel is secreted a little stone corral, shaped like a finger-stall, opening down the gully and blocking off the final exit to the hilltop. A good deal of the stonework has crumbled and grassed over, but it stands there, I imagine to this day, much as it has since the 12th century: an ancient goosefold.

Between July and August, after the goslings have hatched, the pinkfooted geese of the Þjórsárver moult their feathers and replace their plumage. For much of this time, the birds are completely flightless. While most bird species during this period of moult lose only a few of their main flight feathers at any one time, losing more when the first to be shed have been replaced and thus always retaining good powers of flight, the Icelandic pinkfoot loses all its flight feathers in one go, and is flightless until they regrow. Although one might presume this is a risky business, the geese are by no means as vulnerable as this might suggest – they can show a remarkable turn of speed on the ground, outpacing most potential predators, including humans. However, they make one miscalculation. When threatened by a predator in this flightless state, while other geese take to water, the Icelandic pinkfeet run to the nearest hill or crest in the moraine and move gradually up to the brow. Presumably such a behaviour – which I believe is peculiar to the Icelandic geese – has advantages in that it enables the birds to command a better view of the surrounding meadow, better to keep an eye on the approaching danger and, also, in that when towards the end of the moult the birds are nearly flighted once more, they stand a better chance of getting airborne if they launch themselves from a hilltop. But the 12th-century Icelanders were obviously keen observers, and more than able to turn their knowledge of natural history to their own culinary advantage. In natural gullies such as the one on Nautalda, or in depressions just over the brow of other raised areas in the meadow, they built their goosefolds. Then, each July they returned to the Þjórsárver, circled the meadow on their sure-footed ponies and gently herded the geese towards the nearest hilltop or hummock. Over the top and into the pens, and the helpless geese could be slaughtered in their thousands.

The slaughter has stopped and the geese are protected by law; but the goosefold remains and the geese still run uphill. And, with special permission granted by the authorities in Reykjavík, we intended to exploit that very same behaviour to catch ourselves some geese to take blood samples for analysis. We could not use the original 12th-century folds, but we had yards of netting and firm metal posts with which to make our own funnels and pens on the hillsides. We hadn't got ponies, but we could quarter the meadow on foot to herd the pinkfeet in that way. But we would need help; the more bodies we could muster, the larger we could make the driving circle and the smaller the gaps through which the geese could break back through our line.

The camp on Nautalda had swelled enormously. Now in addition to our own tents were those of the Radley College party whose help we'd been promised. We had moved them in the previous night: their Land Rovers had struggled over the moraine and the lava, laden with their own gear and our catching equipment. The 'rovers now rested on the other side of the Hnifá; their drivers had jibbed at the swollen waters and we'd elected to do the last stage on foot. After all we were used to it by now, and with eight more extra bodies to share the loads the packs felt unusually light. Now in the familiar drizzle of the early morning we prepared to move off on our first catch.

We rigged up our nets in a dip just below the crest of a little moraine hillock, Nauthagi, and moved out towards the edges of the area of meadow which immediately surrounded it. To get anywhere in the Þjórsárver from Nautalda involved wading thigh-deep through two fast-flowing streams below camp (the penalty for camping by a reliable source of fresh water). Thus, by the time I had scrambled up onto the moraine below the Hofsjökull icecap, I was already soaking wet (again). The drizzle and mist settled into a more persistent steady rain as I swung up onto the ice and prepared to cross the outfall of the glacier. I'd chosen to go up onto the ice to keep above the meltstream it disgorged, and at least reduce if not eliminate the number of soakings I would receive in criss-crossing the river. I needn't have bothered: my donkey jacket dripped soggily onto the ice as I leapt from footing to footing on the glacier. Finally I reckoned I was round far enough, and slid back down onto the meadow below, Far away, through my binoculars, I could see the others skirting around the edge of the sphagnum, gradually encircling the area.

I scanned the meadow: a seemingly endless expanse of sedge, sphagnum and standing water. Not a goose in sight. I waited, crouched against the rain while the minutes ticked interminably by. Then it was time to close in and we all began to move slowly across the bog, ankle-deep in the murky water, sinking into the spongy moss with every slow, deliberate pace. We moved painfully slowly –every step a slow-motion caricature of a limb in movement. We mustn't panic the geese, but must drive them gently forward between us, barely influencing their movement, just deliberately easing them forward. The rain dripped persistently from my hair and eyebrows. I timed our slow progress. In half an hour I had covered some 100 yards – barely moved from the shelter of the moraine. These Icelanders maybe knew a thing or two about doing it on ponyback. We moved on – and still no geese.

At last the circle began to close – and we began to spy the geese, moving ahead of us between the moss banks: alert, with only the chocolate heads showing above the sedges, or grazing as they went, steadily, undisturbed. Ones and twos, no more. A party of perhaps five broke back between our lines and escaped to the meadow behind. All the geese were beginning to become a little more obviously aware of our presence. Every so often we would come across one shamming dead: adults and goslings alike – flat down on the ground, snuggled into a little depression, neck stretched straight and rigidly forward, eyes firmly shut. They lie so motionless they are incredibly difficult to see, and it is only when you stand right over them and they know you must have seen them that they scramble to their feet and hurry away. Still the bulk of them moved on ahead; I guessed that we had between 50 and 60 geese up beyond us in the meadow.

We closed right in and started to climb the slopes of Nauthagi. Suddenly we could *hear* them: a swelling babble as we scrambled up the moraine, louder and louder, till we were over the shoulder of the hill and could see them on the crest ahead of us. Not 50 or 60 geese as we had imagined, but literally hundreds of them. milling uneasily in front of our nets – crammed together so that there was no room between them, a moving sea of geese, calling incessantly. The noise, and the sight, separately were stirring, together: breathtaking. If ever there was a time

I could wish for the 'feelies' it is in trying to describe that spectacle: a massed, moving tide of beautiful wild geese silhouetted on the skyline of the moraine ridge; the clear babbling call of geese across the wide, desolate meadow. It was absolutely bewitching. Cramped and rain-weary limbs forgotten, we herded the geese down into the capture pen. We lost a good many of them because the capture pen was simply far too small – we had never anticipated catching this many geese in one sweep of the meadow. But when we finally closed the net, we still had some 500–600 birds within the netted area.

We stared in fascination: these ancient Icelanders had obviously been on to a good thing. When Peter Scott had come out on an expedition here with James Fisher in 1951, catching pinkfeet in order to ring them, they had chased the flightless geese across the meadow on ponyback, one or two at a time, catching them by hand. In six weeks they pushed their total slowly up to 1,000 geese; we'd nearly made that in our first catch. Still, we couldn't hang about in self-congratulation: we shouldn't hold the geese for too long in the cold and wet or we'd lose some.

And so the sampling began. Pulling a goose out of the net, holding it akimbo across one's chest. The measuring and the blood-sampling: the needle in the crook of the wing, the slow drawing of the plunger in the hypodermic, the blood welling into the syringe. One millilitre. The flick of the needle, the antiseptic swab clamped to the wound for a minute or so to stop the bleeding. Across to a separate holding area to await release and on to the next one. We worked in pairs, one holding, one sampling, moving soggily in wet socks and soaking boots. We began to move like zombies, pre-programmed for months for this particular task. And still it rained.

An hour passed. The little phials of preserved blood mounted up in the coolbox: 40, 45, 50 samples. The geese were quieter now – and we worked on. But the little pinkfeet were beginning to show signs of the cold, ruffling feathers and settling sleepily down on the rough shingle of the moraine. Silly really, because high on adrenaline, we felt fine. Indeed, we were beginning to feel comfortably warm, a little drowsy to be sure, but perhaps a short rest, – the danger signals. And if we were showing the first signs of hypothermia, what about the little geese?

Hastily we finished sampling the geese we had in our arms, and opened the nets. We'd not done badly anyway for one day's work. The geese stayed where they were, confused, reluctant to go. Slowly they began to drift out in ones and two until, suddenly, they caught the idea and were away, streaming out of the nets, over the rise and away down across the meadow. We packed up our own gear and moved determinedly and briskly back to camp. One thing we didn't want was a serious case of exposure out here, 400 miles and a good 18 hours from the nearest hospital.

We shot into our sleeping bags like rats down a drain and slowly began to warm up. A whisky – the worst thing, of course, for hypothermia: dilating the blood vessels and causing even faster heat loss from an already chilling body – but we felt safe now that we were inside our thick quilted sleeping bags, and we started to relax. It had been a tremendous day: the expedition a success in one fell swoop. We rested happily; weary but satisfied. At last it seemed we were winning.

The following day saw a repeat performance, but this time we were far more confident in what we were doing and far less concerned about losing the odd one or two geese which might break back through the line. We moved in faster and were content to close the nets on the first two or three hundred geese that walked in. Then it was time for the Radleians to pack up and go – taking with them all the blood samples we had collected so far straight down to Reykjavík, to be flown home under refrigeration. Tony and Harry were to leave with them, for they had to return home earlier than the rest of us, and now that we'd achieved what we had set out to do it seemed an appropriate time for them to go. The rest of us would stay behind and try to increase the tally of blood samples.

Coda

There was a distressing sequel to this trip. First, the blood samples we had so painstakingly gathered and returned to England under refrigeration had reached Oxford successfully and were stored in a deep freeze awaiting our return. But two days before we got back to them, the deep freeze faltered in a power cut and the samples were destroyed.

Second, much of the behavioural data had been recorded on film: 1,800 feet of colour film provided by the BBC and a further 700 feet of black and white film from the Max Planck Institute at Seewiesen, courtesy of Professor Konrad Lorenz, whose interest in goose behaviour is well known. An inexperienced technician confused the clearly marked cans and processed both colour and black and white films in the same developer. The emulsions mixed, and the expedition's film was ruined. Little could be salvaged, and from a gruelling four-month trip all we retained were our field notebooks.

The double tragedy struck a sickening blow after all the many tribulations of the trip, but as it turned out all was not lost. Through efforts beside our own (the Wildfowl Trust, World Wildlife Fund and Royal Society for the Protection of Birds), the plan to flood the Þjórsárver was shelved and the meadows remain a beautiful wild space. The Icelandic Natural History Museum, which sponsors all scientific research to be carried out within the island, established a permanent research station on Nautalda; access is easier than it used to be, and a flood of valuable information on the birds of the Þjórsárver is being produced by relays of visiting scientists. In 1990 the Þjórsárver was listed as a Ramsar site (a wetland of international importance), and in 2017 the protected area was quadrupled in size, to 1,563 square kilometres (603 square miles).

It may not have ended as we intended, but the expedition was in its own way a success.

Iceland

Gullfoss, the golden waterfall

Arrival at Kerlingarfjöll

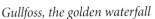

Herding the flightless
geese towards the nets

Golden plover
chicks, Granunes

Moss rings, Loðmundur

East Africa

Jacaranda blossoms and Kikuyu, Kianyaga

Beisa oryx in boma, Galana Ranch

Lions resting in the Gol Kopjes, Serengeti

Lioness with wildebeest kill, Serengeti

Foraging eland

Image credit: Jackie Pringle

Giant groundsel, Teleki Valley

Nigeria

The walled city of Kano

Modern Kano

The author is honoured with a gourd of
yoghourt in a Fulani village

Street market, Nguru

Prayer-bead seller,
Maiduguri

The Maiduguri durbah

The field house, Baluran

The extinct volcano
Gunung Baluran at sunset

The coastal swamp forest

Monitor lizard in the coastal forest

Indian muntjak, Baluran ... but they never stopped long enough for us to dart them

Water buffalo wallow, Bama

*Long-tailed macaques
drinking, Bama*

BOOK 2

African horizons

One

ARRIVAL IN EAST AFRICA

Seven years and a lifetime later, I was standing in the narrow doorway of a 'bucket shop' in the Tottenham Court Road (no internet in those days), booking my flights to East Africa with the delightfully styled Indo-Afric Airways. (I guess the name should have given me a clue, but I didn't make the connection that this particular bucket shop was aimed primarily at non-Europeans; ours were, in the event, the only white faces on board this extremely agreeable flight of Africans and African Asians returning home.)

By this time I had married, and my father had kindly volunteered to get up at some unspeakable hour to act as our taxi to Heathrow, since there would be no public transport available. (It *was* a cheap flight.) Some work I had been engaged upon had finished early and, with a convenient three-month break before I took up my next posting, my wife Morag and I had accepted the invitation of various biologist friends in East Africa to spend the time out there to act as unpaid assistants in some of their ongoing work – and for me personally, to satisfy my curiosity with a visit to the African Wildlife Leadership Foundation's innovative wildlife ranching experiment on the Galana River.

The flight was excellent, if sometimes somewhat cramping, with large African gentlemen in flowing robes on either side overspilling somewhat their own allotted seating spaces. On arrival in Nairobi, the heat hit at once as we took our places in the patient queue for Immigration and Customs. We had been warned about importing apples: Kenya was fiercely protective of its own developing horticultural industry, forbidding the import of luscious European fruits in a bid to force the purchase and consumption of their own wizened equivalents; apples appear on the list of prohibited imports, neatly sandwiched between firearms and ammunition and closely before rat poison. But our 'preserved botanical specimens' went through without a second glance. Our primary hosts, Lesley and Nigel Lloyd, had a homesick yearning for sloe gin, and the kilos of sloes we had sealed within polythene sachets filled with absolute alcohol were already beginning the glorious process of transmutation as the colours and rich flavours leached out into the surrounding liquid.

Lesley picked us up at the airport. Somewhat to our surprise she actually appeared beside us *before* Immigration, but it appeared once again that official niceties were somewhat relaxed here in Kenya.

She whisked us across the busy, blaring town for a drink and a chance to catch our breath at the Thorn Tree bar of the New Stanley Hotel. At that time at least, the New Stanley was the hotel favoured by richer tourists and business travellers. It was also distinguished by being the place where many, if not most, of Nairobi's prostitutes foregathered to ply their trade: all boasting, slung around their necks, the official plastic identity cards vouching for the regular mandatory health inspection. All this in rather brash contrast to the more reserved and more refined Norfolk, harder to find at the opposite side of town, frequented by the cognoscenti of the expatriate community on their visits to town from stations upcountry. We too preferred the more discreet Norfolk, but on this first day the Thorn Tree provided a better place to watch the colour and bustle of Nairobi.

We didn't linger there, for it was a long drive up to the little township where Lesley and Nigel were based. Nigel was a graduate engineer who worked for one of the larger companies contracted to build roads and push other infrastructure into the interior, and was in charge of a sizeable local workforce – and a remarkable number of Irish! Lesley had gone out to Africa as a volunteer with Voluntary Service Overseas (VSO) and was now teaching in one of the local primary schools. She guided her venerable Renault 4 through the outskirts and then – almost immediately, just as in Iceland – we left the tarmac behind and were driving on the hard-packed red dirt characteristic of the majority of the roads not yet taken in hand by Nigel and his team. We trailed behind us the cloud of red dust characteristic of the dry season; the car's windows were tightly shut, despite the heat, in an attempt to minimise, at least, the amount of silt which penetrated and settled inside the car itself.

Those first impressions of a country always linger – and all along the road, small villages and their associated 'shambas', where the villagers grew their few straggly stems of maize, were interspersed with larger plantations of coffee. Africans cycled past at the road edge, sometimes two or three to a bicycle, with panniers piled high with their belongings, but there was also plenty of motorised traffic. Vehicles belonging to western expatriates or contractors were mostly immediately recognisable, for few Africans possessed cars as modern as these, but an abundance of older, somewhat ramshackle vehicles, often clearly hybridised vehicles assembled from the remains of others, weaved their way giddily in and out of the traffic. The majority of these were local taxis, crammed with humanity to the roof – and usually beyond, for it was clearly important not to waste space nor leave any potential passenger behind. There appeared to be few rules of the road. Lesley swerved the Renault violently off the road as we crested a slight rise to be faced with two juggernauts side by side struggling up the incline of the other side, one trying to overtake and the other clearly not willing to give way. She seemed remarkably unconcerned as we avoided an overladen bicycle and swung back onto the harder surface some yards later, simply observing that most Africans were truly fatalistic and worked on the principle that if it was your time to die, it was your time to die; she and Nigel between them voted an 'Idiot of the Week' award for the most outrageous driving they encountered. A big black Mercedes swished past us at speed, driven by two Africans, as if to give the lie that only westerners could afford fancy vehicles; some 20 minutes we saw it again, parked neatly on its radiator grille with the roof and boot leaning gently against a telegraph pole – immediately and unanimously awarded idiot for *this* week.

Eventually, we left the main carriageway for a series of more minor roads, and pulled up outside a neat bungalow under a red pantiled roof. Jacaranda trees dangled racemes of lavender, and the small garden was a riot of colour; a passion vine covered with wrinkled purple fruits hugged a trellis by the door. And the luscious creamy blooms of frangipani scattered their fragrance. After the dust of the journey, this seemed a wonderful oasis – as did the cold Tusker beers Nigel had waiting for us on the porch. Through the trees, and in the distance, glittered the snowcapped peaks of Mount Kenya. The house, in the Laikipia district of Kenya, was Lesley's – or at least had been allocated for her use by VSO – and was part of a wider compound of VSO and other volunteer workers. While clearly full of westerners and not integrated into the wider community, it was actually an open and friendly little enclave: neighbours Ben and Vicky, working for the American Peace Corps and seeing the new faces of new visitors, joined us on the veranda for beers and then invited us all back to their house for a meal – prepared of course in their absence, for most of the expats had 'house boys'.

For the next few days we stayed around the house or took short excursions into the immediately surrounding countryside, for Nigel had to work; but he had booked some days leave in the following fortnight and – generous hosts that he and Lesley were – they were going to take us on safari. In the interim, it was a peaceful time. The run-up at the end of my last project had been somewhat frenetic, and it was nice to slow down for a few days and watch the black-headed weaver birds chattering like sparrows around their cluster of pendulous nests in one of the larger trees within the compound; always busy, tremendously noisy. I have a considerable affection for these birds, which we encountered all over East Africa. True commensals, like our own house sparrows, they always seem to establish their colonies in association with humans: few African villages are without a colony or two in the taller trees in the centre of the cluster of huts, or somewhere on the outskirts. There is a sense of constant activity around these colonies as the birds bicker and squabble between the nests – and there is always a constant coming and going as little flocks leave the nest tree to go out and forage in the surrounding fields, or other groups return. Like avian locusts, large colonies can cause significant damage to crops – especially subsistence crops of maize or mealies – but the birds always seemed remarkably well tolerated by the villagers.

Besides watching the weavers, we had our homework to do; to introduce us to East Africa and something of its history, Lesley and Nigel had set us reading tasks: Alan Moorhead's *The Blue Nile* and Charles Miller's *Lunatic Express*.

Two

MASAI MARA TO
THE SERENGETI PLAINS

Working on the roads, Nigel had unlimited use of a long wheelbase works Land Rover. That weekend we loaded it with camping equipment, food and supplies. Bags of purple passionfruit went in from the vine outside the house, since these keep their moisture and citrus freshness within the withered skin and are an invaluable and refreshing source of liquid. In special carriers lashed to the outside of the vehicle we stowed jerry-cans of extra fuel; there would be times when we were far away from any convenient filling station. Nigel had planned the stores list and the packing to a nicety; a cheerful and easy-going man, he was also extraordinarily practical and efficient. Morag and I flitted around the edges, 'helping'. At last it was time to go.

You are unlikely to need me to describe Africa or its wildlife – all is so accessible in countless television documentaries – but in the flesh, as it were, it is the sheer scale of it that gives pause. The high, sheer walls of the great Rift Valley towering above the level floor that simply seemed as if a chunk of the earth's crust had parted company simultaneously with the land on both sides and dropped away to come to rest some 3,000 metres below; Lake Naivasha, pink with flamingos – and mile after mile of the rift floor itself, with stiff-legged giraffe browsing among the flat-topped acacias, baobabs looking like nothing so much as a rain-bloated tree plucked by some giant from the earth and stuck back down again, roots uppermost. And mile after empty mile of it.

That first night we found a sheltered spot in an oxbow curve of the river and pitched the tents. The air was full of what was to us at that stage unknown calls of birds; francolins scuttled past in the dust. We cooked a meal and sat around the tents with mugs of coffee. And as the light started to fail and the sky reddened, a small family party of elephants came slowly along the bank on the opposite side, silhouetted against the sunset, and came peacefully down to the water directly in front of the tents to drink. In a lifetime of wonderful encounters with wildlife, it was truly a magical moment.

The following morning the elephants had gone and we headed south: our objective the now famous Masai Mara reserve and the Mara river. The rains were not far away, and we planned to follow the migration of the wildebeest back down from these comparative uplands to the grass plains of the Serengeti. Each generation recalls how the numbers of wildlife have dwindled since the old days, and it is of course true that overexploitation and disease have decimated the vast herds of the early colonial era. But in those days, 40 years ago, wildlife still teemed, and we were captivated. In the wooded areas of Mara we encountered endless groups of delicate impala, tails flicking continuously (as do those of fallow deer), groups of topi with their smoky faces at the woodland edge, and the graceful, elegant gerenuk, raising themselves almost effortlessly onto their back legs to browse among the lower branches of the trees, out of reach to more earthbound creatures. Territorial pairs of dik dik at regular intervals kept possession of their little slice of Africa (as grey herons seem to space themselves along an English riverbank). In the scrubbier areas there were massed herds of buffalo, and we regularly saw black rhinos, whether solitarily or in mother-and-calf pairs. Each waterhole had its coterie of bushbuck and waterbuck, and out on the plains proper were warthog, zebra, Grant's gazelle and great herds of the charming little Thompson's gazelle. And away in the middle distance, the wildebeest were massing.

Restrictions were few in those days, and we drove out across the plain to join the herds, following slowly as group after group joined together and started to move more purposefully towards the south and Tanzania. Small groups of zebra mingled with them and seemed to be part of this general drift to the south. Both zebra and wildebeest seemed completely unconcerned by our vehicle amongst them, while for us there was something completely surreal about being surrounded by the slow-moving massing herds on either side, mewling and grunting as they walked. Our cameras clicked busily. Rather than miss a shot we had got into the habit of taking a picture early on, in case we never got any closer to a given subject, replacing it with a succession of subsequent shots if one was able to approach more closely. Here there was no need, for the animals were all around us. Nigel switched off the engine. By now the herds were jostling the vehicle, and pressed so close around us that they moved the Land Rover along with them for quite some distance with their own pressure around it. We would have to break away at some point, for they were headed for a river crossing too steep for us to negotiate, but for now we were completely awestruck by being part of the spectacle.

We, too were headed for the Serengeti grasslands. Eventually we did have to break away from the river of grunting wildebeest and make our own way towards the Mara river, following it down into the Serengeti proper. (I do not recall any border controls as we crossed from Kenya into Tanzania, although we did pass Customs on the return journey on metalled roads.) Although our progress was leisurely and partly in vacational mood, I had already agreed to meet up with a former colleague at the Serengeti research station in Seronera. I had known David Bygott from Cambridge, when, in my early days as a post-graduate, I had worked for a while in the university's Sub-department of Animal Behaviour. During that period, Dave had returned from working on chimpanzees in the Gombe Reserve under Jane Goodall, to write his doctoral thesis. I well remember the occasion when he prevailed upon me to attach a towing rope to my battered Triumph Herald and tow his two-wheeled Norton Commando (with him securely astride it) into the garage in Cambridge for repair. David and Jeanette were now working on lions, and had invited us to come and spend time with them. Additional hands would also be welcome for a proposed capture operation to help a colleague put radio collars on a sample of giraffe …

We found a suitable place to camp not far from the research station, and the following morning joined David and Jeanette as they drove us out to introduce us to 'their' lions. It may seem odd now that we could camp where we chose and did not have to stick to designated campsites. We would perhaps have been accorded special privilege as guests of the Research Institute but, as in the Masai Mara, restrictions were few (indeed, throughout Tanzania); only weeks later we were camped happily in the bottom of the Ngorongoro crater, watching herds of eland grazing slowly past us – and sung to sleep by the eldritch howls of hyenas in the middle distance. But I digress.

Special status did allow us much closer contact with the lions, and a much closer approach than would have been permitted elsewhere. We had of course already seen groups of lions on our travels thus far and seen the crowds of zebra-striped tourist buses clustering round a kill. But this was something different. Because the lions were habituated to David and Jeanette and their vehicle, we could come very close indeed and observe their behaviour at very close-hand.

By now the sun was well up and the day was heating; the rains were not here yet. David drove out to one of the many scattered piles of rocks (or 'kopjes') which form huge cairns across the plain, to introduce us to the Seronera pride. They had obviously fed the night before and all were stretched out on the flatter slabs of rock, resting and sunning themselves. Tails twitched, but no one was going to be doing much. Around the base of the kopje the dry savannah shimmered into the distance in all directions with the classic scattering of flat-topped acacia trees. There was little grass as yet in this end of the dry season, but the plains were peppered with thorn scrub. Startlingly red cardinals flicked among the thorn trees; hammerkop and marabou storks, like elderly militiamen, stalked the dry ground looking for lizards or unwary rodents. High overhead, groups of vultures circled in the thermals, scanning, with their incredible eyesight, for fresh carcases – perhaps searching for the Seronera pride's kill of the previous night.

By noon the sun was really hot, and we soon came across another pride resting in the shade beneath a large acacia. Again, this pride was simply resting – and totally unconcerned by our presence. Six lionesses lounged in patches of shade, accompanied by four or five cubs already a good few weeks old. A single large male, gold-ruffed, dozed amongst them; he must have been their father (or at least have presumed he was) for adult lions usually kill unrelated cubs in order to bring the females back into oestrus as quickly as possible. As the sun moved slowly across the sky, at times he would stand and stretch, swishing his tail against the gathering flies before flopping down once again into a better patch of shade. Only the cubs were active, playing around the base of the tree, pouncing on the twitching tails of their mother and aunts, occasionally demanding that their mother roll over and allow them to suckle. All the group looked sleek and well-fed. We sat less than a dozen yards away with all the windows down, watching, not intruding, and it was an enormous privilege to be allowed to be a part of such an intimate encounter – sharing something which no wildlife documentary can ever communicate, and an intimacy that the cut-and-run activities of the tourist buses could never deliver.

That is because for many tourists, their experiences are those offered by those safari buses whose owners' main aim is to ensure that paying customers get to see as much wildlife as possible in the shortest of time. Eco-tourism is indeed a valuable contribution to the economy, if it can be achieved without excessive disturbance of the animals – but the black-and-white-striped combi vans cluster around and disturb animals at any kill, converge on water holes and often drive away the animals gathered there, and regularly interrupt matings by their intrusion too close to a courting pair. David's predecessor studying the lions at Seronera, Brian Bertram, had written a spoof article for, I think, the *East African Wildlife Journal*, describing the behaviour of these tourist buses. He never once revealed the true identity of the 'species', punctiliously referring to them throughout the article by their Latin binomial *Microbus quadrorotarius* (four-wheeled minibus) – but he described, as might an observant naturalist, how they aggregated at water holes, that their ecology must be that of scavenger or carrion-feeder, since they never-failingly massed around lion kills as quickly as vultures might assemble. He described their pelage (skin covering), as cleverly mimicking that of the plains zebra – and even argued that they must be seasonal breeders, since the cohort of juveniles (identified of course by a change of letter in the registration plate) appeared in synchrony at the same time each year. (I never saw its publication, but it must have resulted in a rare smile in the panel of reviewers who assess the scientific merit of any submission to a scientific journal before it is accepted for publication – and I do have in my files an equally spurious (published) account of the fictional ecology of dragons ...)

We retreated to the lodge at Seronera. We had planned to spend more time with David and Jeanette and the lions, but plans were disrupted by an all-hands-on-deck call for assistance with the giraffe catchup. This had to be accomplished before the rains began in earnest, and an experienced giraffe catcher (there are such things?), travelling up from South Africa to advise and coordinate the catching, had arrived the previous evening. All other work would be put on hold for the time being.

In my professional career, I have caught and handled literally thousands of wildlife individuals – chiefly deer of various species – but that was as yet in the future, and I had certainly never been hands-on to a giraffe. The day dawned dry and clear, the rains still a way off, and we gathered for a briefing in a dry riverbed, wadi, for briefing. Giraffe had been located in a patch of acacia not far away, and the South African expert was going to try to shoot a tranquillising dart into one of them. The idea was not to knock it out completely, but to slow its movements sufficiently that we could subsequently catch up with it and bring it down with ropes. We were assigned to the rope party, charged with getting ahead of the slowing creature, stretching ropes across its path to slow it further, and then to wrap the ropes further around its legs like a living *bolas*. Once it hit the deck, we were told, we would have to act swiftly to lift the head and neck and keep them upright, lest the animal should regurgitate as its chest struck the ground and then choke in its own vomit.

That was the theory anyway – and it worked surprisingly well. As we waited and watched, a black rhino exploded out of the thorn scrub of the wadi, doubtless disturbed by the stalking party. Before very long a large male giraffe approached us slowly and rather clumsily along the dry river bed, shaking its head rather muzzily. While we were ready with the ropes, its shambling forward progress was already hampered by the unexpected presence of the South African expert who, having fired the tranquillising syringe, had set aside his rifle and leapt to catch the tail of the giraffe as it passed him, hanging on grimly to slow it further. We quickly braced ropes across the front of its chest, and he gradually pulled it backwards as its legs buckled under it. The head miraculously stayed up and we quickly dropped our ropes to go and keep it that way. Despite the legendary length of their neck, giraffes have no more vertebrae in it than any other mammal, but I confess I was taken by surprise by the sheer weight of it. Head and neck together were an incredible weight, and we were heartily relieved when the two ends of the hessian transmitter collar were riveted together around the neck and the giraffe was resuscitated with the reversal agent for the narcotics in the tranquilliser. He lumbered to his feet and loped off, apparently none the worse. In all, a further five giraffes were darted, and I think only one did not recover from the anaesthetic.

But this time we could not linger; Nigel's leave was nearly ended, and he had one more stop he wanted to make before we had to return, by slightly more direct routes, to Kenya. We left in late afternoon sunshine, thunderheads building to the west, driving slowly back down the winding course of the river. Hippo lounged in the deeper pools, and we spotted a number of crocodiles on the muddy margins. At this stage of the dry season, the river was very low, and both hippos and crocodile were congregated in the deepest pools remaining. Meanwhile, we were headed for the small Tanzanian town of Arusha with its more permanent water and the lushness that it brings.

Three

ARUSHA

For me at that time, Arusha encapsulated the colours and smells of tropical Africa. Set largely on either side of the highway passing through, it was riotous with colour. While the ubiquitous jacarandas and other tropical blooms made their contribution among the varied colours of the trading stalls by the roadside, the palm-thatched roofs, this was completely overshadowed by the shimmering kaleidoscope of peacock-bright colour of the people's costume and the clothes. The smell, too, of rotting fruit and vegetables (for this was market day and over-ripe fruits do not last long in these temperatures), animal dung from cattle and the many pack donkeys – and the smell of untreated human waste, since sewage ran through open drains throughout the town.

Eye-watering though it might be from the ammonia, eye-dazzling from the ever-moving bustle of exuberant colour in the strong sunlight, I loved it. Street stalls sold cloth of these brighter hues, but also more subdued weave for men's suits or shirts. Under the shade of open fronted palm-roofed booths, serried ranks of busy Singer sewing machines proclaimed the tailors ready and willing to turn such cloth, at once, into the aforementioned shirts or suiting. In the main market place, mounds of fruit and vegetables were piled carefully on raffia mats on the ground, adding their own confusing array of colours: mangoes, pulses, coffee beans, peppers. There were piles too of hay, grass and other forage for the lowing pens of cattle; the sense of constant movement enhanced by the continuous arrival all the time of new relays of little donkeys, panniers piled high with cut grass. Flies everywhere, drawn by the sewage and the rotting fruit – but I absolutely loved it for its sheer vitality.

I have worked many times since in the tropics, but to this day, our arrival in Arusha is vivid in my memory. Perhaps, too, because it is green. Nestled under the slopes of Mount Meru (in those days of newfound independence, renamed Socialist Peak), with its cloud-forest and moisture even in the so-called dry season outside the rains, it was a strong contrast from the sere savannah lands from which we had come. In this part of Tanzania there is a cluster of national parks within a relatively small compass: Mount Meru's own national park, Lake Manyara with its elephants, tree-climbing lions and rafts of flamingos, arid Tarangire, Ngorongoro – and towering over everything, the snow-capped peak of Kilimanjaro, not a hundred miles distant.

But first: Arusha had its own speciality – a factory making pipes (or better perhaps a manufactory, for these tobacco pipes were all made by hand). Curious to visit, we sought it out and were taken around. And in truth it was extraordinarily impressive: wooden pipes of all designs and styles, some lined within the bowl with meerschaum to cool the smoke; others carved from meerschaum alone and stained with beeswax, some boasting intricate carvings on the outer surface, others with the bowl hand-stitched into snug-fitting sleeves of leather. The skill of the craftsmen was remarkable. I was in those days still a smoker of tobacco and purchased a number of these pipes: some pure briars, one rosewood lined with meerschaum, and another leather-jacketed meerschaum in a Peterson style. Forty years later, I still have it as a souvenir, although sadly, it was abstracted from its rack and subsequently redesigned by a young and toothy Munsterlander puppy. (I might add that the pipe had been such a favourite of mine that this occurrence in itself finally decided me to give up tobacco completely; I have never smoked since.)

We tore ourselves away and drove back through town and into the Mount Meru park. Following a track up the side of the peak itself we drove as far as Nigel's trusty Land Rover was able, and ended up in a small natural clearing among the trees high up on the side of the mountain.

Two bushbuck skipped away as we stopped and pitched our tents in the glorious little glade. (I have written little of our diet, for it consisted mostly of tinned or dried foods while we were travelling, although when we ate out I had become extremely fond of mealies – or 'posho' as they are called in eastern Africa: a porridge made of coarsely ground maize and generally served with a meat gravy.)

We turned in as the sun went down and the clearing darkened. I woke early in the unaccustomed surroundings, and the clearing was still cool with overnight dew; a blue sky glimpsed through the trees promised another clear day ahead – but there was surely something here which I hadn't noticed the night before?

High up in the cloud-forest, bathed in moisture every night, the trees were bearded with lichens and other epiphytes, but dangling from the branches ringing the edge of the clearing in which we had pitched the tents was what appeared to all intents and purposes to be a ring of furry bellpulls, as if we had camped in the chamber of some open air belfry. Dangling, white and furry … ? I was enchanted to resolve this as a watching circle of colobus monkeys who then descended from their perch as the others wakened and calmly processed through camp, the rippling fringes of their silky black and white capes as well groomed as the brushed-out fringes of any show Pomeranian. We breakfasted and drove along the ridge to spy from the crater rim into Ngurdoto. Much smaller and less well known than its more famous neighbour Ngorongoro, Ngurdoto is nonetheless a real jewel. Far below us, as small as Britain's farm models, we could see herds of eland, zebra, gazelle grazing across the emerald-green floor of the volcano, blissfully undisturbed; we watched for a while, in no hurry to leave this magical place. Red colobus chattered from the trees around us as we finally turned away and descended back into the bustle of Arusha. For a wildlife experience, that night had to score pretty high in the rankings. And not once within the park had we seen another human being.

As we drove we were frequently hailed or accosted by numerous hawkers or street-traders, who carried their packs or had set up their stalls along this well-frequented tourist route. We were always curious to stop and see what they had to offer, for in truth after the expense of the flights we had little money to spend on ourselves or others back home as souvenirs – and part of the pleasure the traders themselves seemed to derive from the interaction was the inevitable bargaining. Many of the wares on offer were poor: clearly aimed at that same tourist market, these were ill-wrought carvings of African animals – elephant, giraffe, kudu, lion – or mass-produced cloth cut into the traditional *kikoi* (something I myself got into the habit of wearing in the evenings). But on a bend in the road some way out of Arusha I did find something which caught my eye – an exquisite chess set carved of white and black ebony, with the pieces depicting African animals or villages or tribespeople.

Here was the ideal gift to take home to my father as a thank-you for taking us to and from the airport. I thought of the hours of work which must have gone into the pieces, for the carving was intricate and really exceptional, and haggled with only half my heart, for it seemed almost criminal to beat down the price on such hours of work; I think in the end I paid one hundred shillings (about £5 in those days)[4] and felt terribly ashamed …

Overall we had been very impressed by Tanzania. Despite the poverty, the people we met seemed somehow more self-confident, proud, than the Africans we had met in Kenya (especially those in and around Nairobi). These last, despite the token exchange of 'jambos',[5] seemed somewhat cowed and in some ways subdued as if still somehow suppressed. They had cast off the colonial yoke of European domination, it sometimes seemed to us, only to have traded that for similar domination by a largely Bantu elite. By contrast, people in Tanzania, while undoubtedly poorer, seemed proud of their independence and what *Uhuru* [Independence] and their beloved President Nyerere had done for them. To my mind at least, they seemed to walk a little straighter, shoulders proudly back, than the somewhat bowed figures of many Kenyans.

One final wildlife spectacle before we had to hasten back across the frontier into Kenya, and yet another national park: Manyara. Famous to me for the work done there on elephants by Ian Douglas-Hamilton – and

---

4 In fairness, 100 shillings was considered at the time a fair, indeed generous, weekly wage for a 'house boy'.

5 This, the traditional Swahili exchange on meeting – however insincere – of: 'Jambo, bwana, habare? Asante sana. Na wewe mwenyewe? Mzuri sana,' which translates approximately as: 'Hello, how are you? Good, thank you. Yourself? Very good.'

on buffalo, by my own colleague from the Netherlands, Herbert Prins – the park is more widely known for its flamingos and tree-climbing lions. It is a relatively small park, stretched along the side of Lake Manyara. In our time, a smart lodge permitted a view from a railed veranda right across the lake, with its pink thousands of flamingos feeding patiently in the shallows, or in flocks of startling scarlet, showing broad black wing stripes, as one group or another lifted from the surface to wheel above the water and settle in some new feeding place. Along the shore was grazing for elephant and buffalo, fringed by a corridor of those curious sausage trees whose long pendulous pods hang down from the canopy on thin stalks like colobus tails. Towards the edge of the park the land lifts into a steeper escarpment, for this is again the Rift Valley, and denser forest cloaks these steeper slopes. Indeed, it was here we saw our first forest elephants – a race or form distinctly smaller than their savannah relatives.

It was in Manyara that the ecologist Herbert Prins demonstrated one of the now classic examples of group decision making among animals, showing that individual members of a group of female African buffalo 'vote' on where the group should next travel to feed. Among buffalo, males and females are separated into single-sex herds (males often tend to occur more solitarily or in much smaller groupings than the larger herds formed by the females, which are often composed of a matriarchal female and her various daughters of different ages). Females move around a predictable home range, and although they will browse they are chiefly grazing animals, moving therefore from one to another of the grassy clearings which punctuate the scrubby woodlands within their range. Clearly some of these are clearings of higher quality than others, but equally the choice of which clearing to visit next must be affected by how recently the herd was last there and how long, therefore, the grasses have had to recover and produce sufficient new growth to be worth returning to.

Like many grazers, buffalo have within the group a synchronised pattern of grazing for a period and then lying down to ruminate and chew the cud, before rising to graze once more. Prins noticed that when they first lay down the buffalo cows might face almost any direction. Occasionally one might get up to stretch or defaecate before lying down in a slightly different position. And he realised that over time the direction faced by all the cows became more and more similar. Eventually, they would all get up and move away – in the exact direction pointed to by the average vector of the final resting position of the different animals within the herd, all making for the nearest grassland clearing within their range that lay in that precise direction. Sometimes, the decision was less clear, but amongst a group of experienced (older) cows, the positions adopted by the different members of the herd at the end of a period of rest was more or less unanimous. Since that time, various other examples have emerged of decision making within groups of animals in this way, but Herbert's work still provides one of the most elegant demonstrations.

But, of course it was also in Manyara that naturalists found a classic example of cultural traditions developing among animals, for there, with no kopjes or rocky areas commanding a good view of the surrounding terrain, lions have taken to resting in trees. Leopards of course regularly climb trees and indeed habitually use forks in the trunk to cache their kills; by contrast lions are not in general good climbers. But sure enough as we drove down through the park, we were to pass beside – and under – numerous lionesses sprawled along low branches, paws and tails dangling to either side. It is reported – although the tale may well be apocryphal, or at least have gathered substance in the telling – that when he was working in the park studying elephants, Ian Douglas-Hamilton had developed for himself a form of entertainment which was based upon driving through the park in his open-topped Land Rover, grabbing the tails of sleeping lions to give them a strong tug, as he roared away underneath…. (The game had the added spice, of course, of the risk that he might simply dislodge them from their tree to fall into the back of the vehicle itself.)

I assure you, stumbling upon buffalo in thick scrub, well within their personal distance is excitement enough for me without *inviting* trouble.

Four

GAME RANCH AT GALANA

Nigel's leave was over and we had to return to Kenya, for Lesley too had to start a new term at the local school. As hosts they had been generous to a fault (although we may hope that they had enjoyed the trip too!), but they now surprised us still further by offering us the loan of Lesley's valiant Renault 4 to drive down to the coast to Mombasa and Malindi and then up to the Galana Ranch. I had hoped to visit colleagues there, exploring the potential of ranching semi-domesticated wildlife species (or at least tamed individuals) in semi-arid or arid areas as potentially a more productive use of land than trying to herd established domesticates such as cattle, sheep and goats. This was of especial interest to me as a biologist, since I myself had been becoming increasingly involved in early experiments in the UK with deer farming.

The project at the Galana Ranch had been established and funded by the African Wildlife Leadership Foundation on the principle that animals which had evolved in the surrounding landscapes would be likely to have developed a number of natural characteristics which enabled them to thrive and to deal with its challenges more efficiently than introduced domestic livestock, even if these had been developed in Africa. The essence of the scheme was that the wild ungulates should be habituated to human contact so that they could be herded through the bush by native herders in just the same way as they might herd sheep or goats, and thus returned to the safety of a compound, or 'boma', overnight to protect them from predators. The thought was that native ungulates might have a greater tolerance to heat than domestic livestock (and waste less energy in thermoregulation), might also be able to exploit a wider range of foodstuffs (having adaptations to deal with thorny or otherwise unpalatable species), with in consequence a more diffuse and diluted impact on the vegetation itself than their more specialised domestic counterparts, and finally might well have greater resistance to disease. In arid areas, too, cattle (for example) need to drink at least daily – if not more frequently still – and thus the herder can only exploit areas of grazing half a day's walking from permanent water; by contrast many of the area's native herbivores can go for many days without drinking (gaining such water as they require from the vegetation they eat), thus greatly increasing the available ground which might be exploited.

All very good in theory … and the Galana Ranch had been established as an experimental facility to test it in practice. Could oryx, eland and buffalo become sufficiently habituated to human contact to allow themselves to be herded alongside or instead of the traditional cattle, sheep and goats? Would they exploit a wider range of forages? Would they prove less susceptible to the heat, or long periods without water? Would they prove more resistant than domestic livestock to local diseases? In summary, would they have a lower environmental impact than the livestock traditionally used in this area and – in using up less energy simply to combat the heat and aridity – would they be able to direct more of the energy they consumed in their foraging into production of protein, with faster growth rates and faster reproductive rates?

At Galana, sizeable collections of fringe-eared oryx and the lovely, elegant eland antelope had been established, as well as a much smaller, and very much more experimental, herd of buffalo. (Ultimately, the buffalo proved too unpredictable in temperament and were dropped from the trials.) Each day, herders were employed to take small groups of oryx and eland out to the surrounding bush to graze, sometimes on their own, frequently alongside the more traditional cattle, sheep and goats. If close to the ranch they might return to the compound that night, or alternatively, if at some distance, utilise an existing boma or build a new one out of thorn trees, ready for the night.

Two western postdoctoral scientists, Mark and Jeff, had been brought in to run the programme, caring for the animals, maintaining the facilities and coordinating the research. Every so often, when back in camp, the groups of oryx and eland would be run through a handling facility alongside the goats, the humped Zebu cattle and the fat-tailed sheep peculiar to Africa, to be weighed and blood-sampled. Blood samples were carefully analysed for evident signs of disease or presence of parasites; detailed records were maintained of weight gains or losses, growth rates of juveniles, periods between the birth of calves. It was an impressive programme – and a great deal of hard work – but both Mark and Jeff had worked in Africa before and had an extremely good rapport with their local staff.

We drove slowly down to the coast, getting used to the curious 'umbrella handle' gear shift characteristic of that model of Renault, stopping only overnight in the increasingly tourist-ridden town of Mombasa, before heading up along the coast to the then much smaller (and to my mind much more attractive) Malindi. Lesley had recently had the Renault serviced, and although I was somewhat apprehensive of what we had already seen of the African's fatalistic attitude to accidents, we had no mishaps. Indeed, for much of the way the road even boasted a surface of tarmac.

Once past Malindi we would have to head inland to find Galana, but Mark had given us clear instructions on the route we should follow. Elephant skulls were commonly used as road markers or mileposts, and we were confidently instructed to turn right at the third such marker skull. Unfortunately, either Mark's counting skills were more limited than I had imagined or some additional elephant had perished in the interim along the roadside: we turned right at the third elephant skull as instructed, and headed further and further into untamed bush but there was clearly no road, and I became increasingly concerned that we had come a much greater distance than we had been told would be necessary.

We dropped down into a dry riverbed and I switched off the engine, to think what to do for the best.

We had no idea where we were, and in truth there was little alternative option but to try to rejoin our main road, travel on to the next elephant skull and see if we fared any better. Retracing our route thus far should not be too difficult, because our fresh tyre tracks still showed in the dirt behind us. I turned the key in the ignition to restart the engine – and it sheared straight across. The handle remained firmly between my finger and thumb while the serious part of the key itself remained equally firmly in the ignition in the off position. Now I *was* worried. As so commonly in Africa, there *was* no spare key. We were we knew not where, miles from anywhere and stuck in a dry riverbed. We could get out and walk … but in which direction and to what purpose? Now, one of the remaining assets of a misspent youth is that I *did* know how to bypass the ignition of a car (I couldn't even attempt it on more modern vehicles) – but as part of its recent service in advance of our trip, Lesley had specifically had the steering lock repaired. While I might start the engine, there was no easy way I was going to be able to release the steering rack without the key.

I don't often panic … but I must admit, things were not looking too promising – when out from the bush rode an African, astride an extremely noisy and very smoky moped. Wearing the remnants of what must at some time in their history have been a pair of shorts and the mandatory flip-flops (open-toed sandals manufactured from an old car tyre), he studied us with interest. Communication wasn't going to be easy since I had accumulated little Swahili beyond the courtesy of greeting – and he clearly had no English. Somehow, however, he seemed to grasp the problem remarkably speedily, produced from the kitbag hanging from the saddle of his moped a pair of needle-nosed pliers and carefully extracted the business end of the key from its imprisonment within the ignition. Wrapping both it and the head of the key carefully in a piece of cloth, he made somehow reassuring gestures (well, I think they were meant to be reassuring), remounted his moped – and disappeared. Well; we were stuck, whatever, so there was little that we could do but wait.

What felt like hours later, but was probably a matter of 50 minutes or so, our good samaritan reappeared and with a flourish presented me with the results of his labours. Somehow (and I have no idea how) he had managed to braze together the two parts of the key into a single whole, filing down and re-profiling the join until it was virtually undetectable. It fitted easily into the ignition and, having been so recently serviced, the Renault purred into life at the first turn. Our angel of mercy smiled broadly, refused all payment, and disappeared into the bush. If he is still alive and remembers, I thank him from the bottom of my heart, because it could have been a very sticky situation. We threaded our way back to the road.

The addition of an extra elephant skull (or at least the potential for such a random addition) was readily explained when after rejoining the main road we came up behind another large group of elephants a few hundred yards further up the main track. There were animals on both sides of the carriageway and they had spilled all over the road. I hooted the horn; they looked curiously towards the car, but made no other move. They were clearly in no hurry and I decided that perhaps neither were we; I shifted the Renault cautiously into reverse gear, but left the engine running, just in case. If that key should break again … (It didn't, and Lesley continued to use it without mishap for the rest of the life of the car.) Eventually, they sauntered off and we carried on up the road to find indeed another elephant skull by the roadside. By contrast to 'ours', this one was freshly painted a clean and sparkling white. In less than half a mile we dropped down towards another streambed to find a small footbridge, and waiting for us on the other side our hosts' Land Rover.

We hastily transhipped our gear, profuse in our apologies for their long wait. But we needn't have worried: the bush telegraph was amazingly efficient and they had heard about our mishap and our eventual release, so had not in fact been waiting long. Cold beer had rarely tasted so good as we sat on Mark's veranda above a bend in the Galana River, sipping chilled Tusker. Below the veranda on the river bank a sign said clearly: 'Beware: Crocodiles'. On the reverse someone had painted in equally clear script 'Beware: People'.

The following day we set out into the bush with one of the African herders. Most of the Galana staff were Samburu, a tall, graceful people – and as skilled with livestock as the more famous Maasai. We were accompanying a mixed group of eland and small stock, chiefly goats. We moved slowly across the dry savannah. Outside the rainy season this was an arid place. What little grass remained was scorched and sere, but the eland plucked at the straw and browsed fresh leaves from the abundant thorn trees. The goats foraged freely around the perimeter of the herd, frequently falling behind and then hurrying to catch up again. While I confess I would not wish to earn my living as a herdsman, it was extremely peaceful. From time to time an eagle would lift from its perch in one of the taller thorn trees – and occasionally, in the distance, we would spot wild oryx.

Perhaps surprisingly, however, we did not see a great deal of wildlife – but this was the end of the dry season and there was little food or water on the plains; wilder creatures might also have been put off by our grazing herd as it rolled slowly across the savannah.

With our limited Swahili we were not able to communicate much with our guide, and Mark had not accompanied us as translator since he was making preparations for a handling session due the following day. But I was enormously impressed with how tame the eland appeared and how easily they responded to the commands of the herder – whether spoken or simply by gesture. Eland are amongst the largest of all the antelope, with males measuring about 6 feet at the shoulder and weighing up to 950 kilos – although females are slightly smaller. But they seemed remarkably docile (and I have read elsewhere that females can become tame enough to be milked).

Without the herds or ourselves being fully aware of it, we toured across the ground in a large, loose circle and returned by late afternoon to the main compound; the goats were detached and the eland herded into a large open enclosure for the night. We strolled around the rest of the facility; gradually other groups were being returned to the compound, and the eland and oryx put into large open enclosures with hurdled walls of woven branches. All seemed remarkably ordered and remarkably calm – a calm to be shattered soon enough the next morning as small groups of animals were herded into the raceways of the covered handling unit to be weighed and measured.

This was indeed a much noisier operation, as each animal was herded into the crush to be weighed and sampled. Mark and Jeff recorded weights and deftly took bloods amidst a constant background murmur of lowing, grunting and bleating from animals registering protest at being so restricted. Each animal, even the smallest of calves, bucked high in the air as it left the crush, delighted to be free at last of the restraint.

We ourselves were sleeping in a permanent tented structure: two hurdle walls woven of branches with a branch and palm-thatch roof. Within this outer wooden construction, open like a tunnel at both ends, an inner tent could be hung when needed, to be removed and folded away when not required. Three or four of these structures had been built in the shade of a large tree behind Mark's own bungalow, and it was actually very comfortable within the inner lining, for it was cool beneath the tree. True, at times we were disturbed a little at night by the sounds of animals crashing around nearby. But it was only on our last morning, as we detached the inner lining on departure, that Mark admitted that in the absence of human occupants, the outer structure of each shelter tended to be used by local buffalo, themselves keen to enjoy the shade of the sheltering tree; the nocturnal crashing was merely the sounds of frustrated beasts finding our own particular enclave already occupied.

Buffalo had not proved sufficiently tractable to develop further as potential alternatives to conventional livestock – indeed, Jeff bore the scars as evidence. But both eland and beisa oryx could be handled, and at the end of the trials it was apparent that both were more efficient food-converters, more drought resistant and more productive under this management system than cattle sheep or goats. They also had a more rapid turnover time, with shorter intervals between calvings, and much faster growth rates of the juveniles. But it seems it never pays simply to consider the biology. At the end of the experiment, when the facility was being dismantled and the animals dispersed, the oryx and eland were offered to the African herders who had worked so long on the project. They refused. Politely Mark asked why. Surely they too, having worked on the project were well aware of the advantages: that these animals could be herded just as easily as sheep or cattle, but were more productive in so many regards. The staff agreed: these animals were productive; they were efficient and the meat was good (we ourselves could vouch for that). However, there was a problem. In their cultural tradition, cattle and sheep and goats were not only their livelihood but also their currency. 'And how can I persuade the father of my wife to take the bride-price in oryx when he is expecting sheep or goats?'

Five

CHRISTMAS ON THE EQUATOR

We had returned to Lesley and Nigel's comfortable bungalow in time for Christmas, to be informed that they felt all wrong about spending the festive season in such heat, and proposed to pack Christmas into rucksacks and spend Christmas Day itself in the snows on the top of Mount Kenya. Well, they were our hosts, so who were we to argue …?

We piled into Nigel's Land Rover again and drove the short distance to the foothills of the mountain, and the best-known (and easiest!) route up. While some of Nigel's colleagues were avid hill-runners and could actually get to the top and back in three hours or less, we ourselves planned to take it more slowly. For a start, we were not yet fully acclimatised to the altitude, and none of us wished to be visited by altitude sickness, to sour the trip.

The vegetation on Mount Kenya shows a clear zonation as you climb. The lower slopes are thickly wooded, but as you gather height this grades into a zone of dense scrub before this in turn is replaced by a region of tundra-like grassland and beyond that an arctic-alpine montane flora peculiar to this part of Africa. We took the ascent slowly; only the previous day an African had been killed on the mountain, stumbling too rapidly on a buffalo concealed in the thick scrub. Buffalo account for more human deaths in Africa than any of the big cats …

We climbed above the trees and cautiously through the thick scrub, making as much noise as we could to alert any sheltering buffalo, and up onto the grasslands. From here the regular route takes you through the Teleki Valley, a curious moonscape of jumbled rocks and tussocky grasses interspersed with the unexpected forms of giant groundsel and giant lobelia. At these altitudes plants can do some curious things, and giant groundsels bear little relation to the humble British groundsel – indeed much more resembling yucca plants; amongst them the tall, straight bushy foxtails of the inflorescences of giant lobelia. The Teleki Valley is named after Count Samuel Teleki, a Hungarian nobleman and explorer who first plotted this route up the mountain towards the end of the 19th century; he is remembered not only in the name of the valley, but in the scientific name of the giant lobelia *Lobelia telekii*.

Onwards and upwards. Progress was slower now, since we were all feeling the effects of a shortage of oxygen in the thinner air. We had no plans to scale the real summits of Batian, but were content to find snow on the top of Point Lenana, a lesser peak. But there was no rush: we would not get to climb to the top this same day, since once the sun is up and the heat has melted the snow and ice, the final scree slopes become too unstable to climb. Instead we planned to spend an uncomfortable night huddled in a hut thoughtfully provided at the base of those final screes, and climb to the summit in the cold early hours after dawn the next morning.

The hut was cramped and uncomfortable; fortunately, we were the only people on the mountain on that occasion, for it would have been *really* crowded had we had to share. We catnapped fitfully, but we got at least some sleep, and the next morning scrambled the last few hundred metres up the frozen scree onto the summit. We spread out oilskins onto the rocks and solemnly handed round the turkey sandwiches.

Coda

It was a fine and fitting end to the trip, for shortly afterwards we had to leave Lesley and Nigel and make our way back to Nairobi and the flight home. That, it must be said, was not as pleasurable or uneventful as had been the flight out, since the Kenya Airways flight was delayed numerous times and eventually cancelled. An aircraft was out of service for unexpected repair, and they had insufficient other planes available to cover all scheduled flights. Because they did not announce this at the outset, but each time promised only a few hours' delay, none of the flight's 150 passengers dared to leave the terminal building, and we spent a most uncomfortable five days sleeping on the airport floors until we all finally rebelled and, in full force an angry deputation, demanded that they take us home. In those days before mobile phones, there was no easy way of getting warnings back home and my father earned his chess set for all the wasted journeys he had made to and from Heathrow.

Six

GO WEST, YOUNG MAN!

Shortly after our return from Kenya, Morag and I had moved home, to the outskirts of the New Forest, and I had taken up a post as lecturer in biology at the University of Southampton. I was happy in Southampton: I enjoyed teaching and had been successful in building up quite a large research group of wildlife biologists. All of our work was applied: primarily focused on studying the behaviour and ecology of large hairy herbivores, as we came to call them, and their interactions with the vegetation, in the interests of applying that new understanding in the development of more sensitive and effective ways of managing them – and their impacts upon their environment.

We had, I like to think, built up quite a good reputation both for research and for the ability it gave us to broker improved and innovative solutions to management problems – but I confess I was nonetheless surprised to receive a call from the British Council asking me if I would be interested to travel as their representative to northern Nigeria to look into a problem of a deterioration of native range due to overgrazing, and suggest possible improvements. Aged 34, I did not feel in any way an expert, and certainly not in the dynamics of African vegetation, although I was assured by my staff that a consultant is simply a man with a briefcase from more than 30 miles away.

To give them full credit, the British Council were tremendously helpful and tremendously supportive. They would arrange my travel, and their resident agent in Kano would meet me from the aeroplane and brief me. I would be assigned a counterpart from the Ministry of Forestry and Livestock, and a driver. All I needed to do was have the necessary inoculations. They even provided me with a modest budget with which to buy suitable clothing (they recommended cotton underpants for the heat). By fortunate circumstance, the cotton safari suits which I had seen to be standard wear during my time in Kenya, with their short-sleeved jackets and short trousers, were all the fashion that year in the UK, so I was able to buy two very nice sets extraordinarily inexpensively in a local outfitters. *Un*fortunately, in my briefings no one had warned me that it isn't polite to show too much bare flesh in a largely Muslim country. The natty safari suits got little wear, and my one pair of lightweight slacks and my one long-sleeved jacket got frequent laundering.

Arrival in Kano was confused and confusing. The intense heat was intensified further by the corrugated iron roof of the little shed into which we were shepherded for the formalities of Immigration. But there was no formality: no polite queues towards uniformed immigration officers in tidy kiosks. Everyone surged forward, jostling to get the attention of the one or two officials seated behind a long trestle table – designed, it appeared, more to keep some distance between them and the surging throng than for the paperwork it might support. It was an insane, disordered, jostling mass of humanity, and I confess I was more than a little concerned when someone beside me – clearly a fellow-passenger – snatched my passport from my hand, and it was passed from hand to hand over the heads of the crowd to the front desk. Clearly immigration was simply a question of ticking off passenger names on some prepared list to ensure that everyone scheduled to arrive had done so, but I was unaccustomed to such mayhem and was heartily relieved when my passport was eventually returned to me by the same overhead route.

Customs was equally cavalier, except for the rigour with which the money I had with me in the local currency and my travellers' cheques were counted and accounted for. There were strict regulations about the amount of money allowed into the country, and even stricter regulations about currency exchange. All must be done at government-approved offices at the official government exchange rate. One was then provided with a stamped certificate confirming the transaction, and these certificates would be checked against your remaining currency on departure. Needless to say, no one actually stuck to this, for exchange rates (particularly of sterling or American dollars) were significantly higher on the black market – and you could buy very realistic-looking certificates of exchange or, if you preferred, the necessary official rubber stamp, on almost any street corner.

The British Council's representative, dark-suited and crisp against my soggy, travel-stained sweatiness, met me at the airport and whisked me away to my lodging. For my time in Kano, I had been allocated a British Council-owned bungalow in a fenced and gated compound down-town. To my amazement, an armed guard sat on a folding chair beside the gate. My guide would meet me in the morning and accompany me on the necessary round of visits to officialdom in the various ministries and research institutes, and then I could take an internal flight further upcountry to Borno State. But for now I was free to explore, although he would encourage me not to stay out too long in the heat of the day when temperatures in the sun could easily get as high as 130–140°F (55–60°C).

Kano was a study in contrasts: originally a fortified city, its walls had remained largely intact: inward-sloping and buttressed walls of mud brick, of a design much as one might see on sea-defences around the British coast. Within these were narrow streets and alleyways between, low houses clearly of Arabian style, also built of mud bricks, cool under the trees. In one quarter, another compound containing the royal palace, for each region of Nigeria still had its own emir who handed out justice and every afternoon would sit to hear the grievances of any complainants who had sought an audience. This inner city might not have changed for centuries; beyond its walls was another world altogether: broad highways, rushing traffic, major interchanges with overhead traffic lights, the onion-shaped domes and towering minarets of a modern Muslim city. Already in the mid-afternoon heat the loudspeaker blare of a distorted *adhan* (Muslim call to prayer) sounded above the noise of the traffic. On the pavements, people were busy chopping and bundling firewood; goats wandered everywhere, within and without the walls, searching for titbits on the many rubbish heaps.

It was a bustling place. Many of the men wore conventional slacks and (to my relief) short-sleeved jackets, but others wore beautifully embroidered cotton tunics, and many the full flowing robes and head-coverings of Arabia, for this was an Arab town and a curious mix of Muslim and Christian. Completely heedless of the advice of my British Council mentor, I explored happily for many hours before returning to my compound and forcing my long-suffering guardian to unlock the gate.

The following morning was indeed the immediately-forgotten round of officials: the Permanent Secretary of the Ministry of Animal and Forestry Resources, the Chief Veterinary Officer and his deputy, the Chief Livestock Officer, the Assistant General Manager of the Chad Basin Development Authority and the Programme Manager of the Borno State Accelerated Development Area Programme ... it all became a bit of a blur, but I find I still have detailed notes in my files, since these were the people to whom I would present my report in person on my return.

I was to travel on their behalf north and east into Borno State (which stretches out towards Lake Chad, bordered to the north by Niger and to the east by Chad itself). Here the various officials were concerned that there were problems with overgrazing – especially in areas with only seasonal rainfall, but even in the uplands with good rains and year-round pastures. Much of the area is grazed by the cattle of migratory herdsmen (who use the seasonal grasslands as part of a wider rotation and typically graze their herds there after the plants have flowered and set their seed, to ensure an annual replenishment of the sward). But the local Kanuri villagers may graze a few sheep and goats on these pastures year round.

Borno, I was informed, consisted of five distinct ethnic groupings. There were the Kanuri, the Hausa, the Shuwa, all Arabs and all resident in permanent villages. There were also the people of Biu, generally Christian (and for some reason my diary tells me, generally well-educated – as if these other ethic groups were not). Finally, there were the Fulani. It was the Fulani Arabs who had played a key role in the 19th-century revival of Islam in Nigeria. While a proportion of the Fulani lived a more settled life in established villages, the majority remained as nomadic herdsmen, moving freely across the borders between Borno, Niger and Chad in their seasonal migrations with their massive herds of cattle. And over those wanderings, and indeed over the numbers of the cattle, Nigerian officials had no control. Many of the Fulani carried arms to protect their herds, and they had the reputation of being forceful and occasionally actively aggressive in defence of their traditional grazing grounds.

I began to appreciate why someone as youthful and inexperienced as myself had been awarded this job by the British Council. No one with more experience would have accepted it in the first place. Fortunately, as I was to discover, my nominated counterpart, Dr Gadzama, the livestock officer for the area, was both experienced and highly intelligent. I was to point out to him on many occasions that he already knew half the answers and was way ahead of me; but he wryly pointed out that a report from an outside 'expert' such as myself would carry far more weight than any internal report he might submit to his seniors: he needed me as much as I needed him.

Briefings over, my British Council minder drove me to the airport. There were few formalities here: a green-liveried Nigerian Airways plane sat on the dirt at the edge of the runway while our pilot lounged against the fuselage, smoking a cigarette. Perhaps the only formality was that passengers were weighed alongside their luggage, in case the combined weight of both might exceed the recommended flying capacity of the aircraft, and then we were boarded. I wedged myself uncomfortably between two clearly affluent and thus well-fed Nigerians in flowing robes, and the aircraft lumbered – rather slowly, I felt, given all the precautions about our collective weight – into the air.

Seven

BORNO STATE

If Kano was hot and dry, Maiduguri was hotter and drier. But I was met swiftly by my appointed colleagues and driven to my hotel. The Deribe Hotel (nowadays, I see, restyled as the Maiduguri International Hotel) was clearly intended to be the New Stanley of Maiduguri: the classy place to be seen, attracting business travellers and Nigerians keen to put on a display of wealth. It was in consequence a somewhat soulless place, but it would have to do. At least an overhead fan worked, gently stirring the tepid air above the bed. I threw my small overnight bag on the bed and went out to explore. My newfound travelling companions, having met me at the airport and escorted me to the hotel had, with total courtesy, left me to my own devices for the evening, promising to collect me in our smart black limousine the following morning. I wandered out of the hotel. Across the road a group of Africans were tinkering with their motorcycles outside the delightfully named God Help Me Patient Medicine Store, advertising pat(i)ent medicine cures for everything from diarrhoea to venereal diseases ... There were many such colourful (and optimistic) advertisements on the shopfronts as I passed along the road and out of town. At the edge of the town stood a permanent street market, operating from dawn to well beyond dusk every day of my stay in Maiduguri, with its stalls piled with fresh fruit, hand tools, bales of colourful cloth. (Towards the end of my stay I bought such a bale for my wife, an accomplished dressmaker – but somehow the weave which had seemed so fitting against the cocoa-black skin of the Nigerian women and under the bright topical sun always seemed too garish against a European complexion under overcast British skies; these things do not transplant effectively.) Beyond the edge of town a scattering of huts and a clustering of 'shambas' – almost like English allotments – with women hoeing the earth between the straggly rows of maize: the 100-day corn, bred to germinate, grow and form cobs in the short period of the rainy season.

I wandered back to the hotel and that lazily spinning fan. Breakfast the following morning (and indeed every morning) was unmemorable and international in the characterless dining room. A country and western tape played quietly from the corner speakers. Kenny Rogers. Every morning. It was as if the management only possessed a single audio tape but were determined to appear modern and up to date by offering musack for their customers over breakfast. Now, I don't dislike the music of Kenny Rogers, but a continuous onslaught does begin to pall. And breakfast in Maiduguri will for me always be associated with the endlessly repeated strains of 'You picked a fine time to leave me, Lucille.'

I was rescued promptly, as promised, by Dr Gadzama, livestock officer, and Alhaji Zanna Talha, field officer and driver (and a very good driver he was). Gadzama, my 'official' counterpart, was a tall well-built man, dressed in an outsize embroidered tunic and loose baggy trousers or pantaloons drawn together at the ankles. He was perhaps a little older than I was myself – and although initially rather formal in his manner towards me, resolved over the course of a journeys, to be a cheerful and easy-going companion. As the livestock officer for the region, he knew the country and its problems well, and set himself early to fill me in with the background. Our driver was an older man, tall and lean in flowing Arab robes; some years earlier he had made the pilgrimage to Mecca and as such was accorded the title 'Alhaji', and entitled to the red fez he habitually wore upon his head. Both were intensely black of skin, perhaps exaggerated by the light-coloured robes they wore, but to someone accustomed to the more varied colours of East Africa, with its darker and lighter chocolate tones ranging to the café-au-lait complexions of the Samburu, the blue-black of Nigeria was striking.

As one who had travelled to Mecca, Field Officer Alhaji Zanna Talha was clearly a devout Muslim, while for his part, Gadzama explained, he was a Christian. Not that it mattered, for, as I have already hinted, this part of Nigeria was an intimate mixture of the two religions side by side with what appeared to be perfect amity. Indeed, Gadzama told me that while he was Christian his brother was a Muslim. (Throughout my stay I experienced no discrimination, and I have to say I found it extraordinarily heartening.)

By contrast, the only other European resident in the Deribe Hotel was an electronic engineer from England working on some agri-industrial development on the outskirts of town. He was somewhat disparaging of the local Nigerian farmers, describing how they would all receive brand new tractors or other machinery with each new wave of international aid. But they did not have the skill, or interest, to maintain or repair them, so that when they broke down, their owners simply left them standing where they were in the fields and awaited the arrival of the next batch of new machines. Bluff and brown-bearded, this man took little interest in much beyond his job; I think I never saw him take a walk through the streets or surrounding countryside: his only relaxation after work appearing to be the consumption of copious quantities of the local beer. Each night he had a different girlfriend – who always seemed to leave before breakfast. The hotel management appeared to turn a blind eye to these additional overnight guests (perhaps they were complicit in their supply) – but I grew increasingly concerned that before too long my young associate might be needing the services of the 'God is Good' medicine store across the road. Although he was in fact an extremely pleasant and friendly fellow, I found his somewhat superficial lifestyle, well-insulated from the wider world around him, rather uncomfortable, accompanied as it was by a rather patronising attitude towards the local people among whom he had been 'forced' to come and work; but it was not, I fear, untypical of the attitudes of a number of expats on shorter-term contracts.

As I have already suggested, the focus of my own project was to explore the issue of overgrazing from domestic livestock in the natural rangelands against the Chad border. These arid areas with very low annual rainfall were grazed by the livestock of the resident tribesmen – Hausa, Kanuri and even those of the Fulani established in more permanent settlements – but also by regular influx of the enormous herds of cattle of the more nomadic Fulani Arabs. Overgrazing had started to change the structure of the rangelands, initially eliminating perennial grasses and herbs in favour of those which could re-establish themselves annually from seed, but ultimately through suppression of more palatable species altogether and – through a reduction of competition from these more vigorous grassy species – facilitating the expansion and proliferation of unpalatable or actually noxious species with better defences against grazing, such as *Cassia tora* or *Acanthospermum paradoxum*. The effect on the fragile vegetation, with the elimination of palatable species and the proliferation, through competitive release, of unpalatable or poisonous thorny scrub species, was to reduce productivity for the grazing herds themselves; worse, the exposure of the dry soils in a windy environment also led to significant soil erosion.

I wasn't sure I would know what to do about the problem, but the first thing was to experience something of the issue first-hand. Our first stop, at Gombole, neatly encapsulated the entire problem. Side by side here at Gombole, as in other parts of Borno, operated two completely different grazing systems. One part of the area was enclosed into a huge cattle ranch of some 20,000 acres (c. 8,000 ha). This one was, I believe, state-owned, although in other regions such cattle ranches might be in private ownership. Within the perimeter fence, a set number of stock was maintained. Gombole used to be subdivided into numerous paddocks, worked separately in rotation, but by the time of my arrival most of those internal fences were in disrepair and animals were herded throughout the ranch, albeit taken to different areas each day. The problem with this sort of setup, however, is that it is enclosed, and so to a large extent the herds are sedentary within a comparatively small area (given the quality of the grazing and the dry conditions for growth). Because of this, even though there may be limited rotation between paddocks, the time interval before their return is short (and grazing pressure is high). In consequence, even during the brief rainy season when everything should be at its best, the grass and palatable forbs (herbaceous flowering plants) are very quickly gone. In the short term of course, this means that the available grazing is used up and there is nothing left to support the herds through the long dry season. In the longer term the overgrazing leads, as above, to the eradication of the more palatable species and their rapid replacement by non-forage species which quickly proliferate (and in turn prevent recolonisation by grasses and palatable herbs. Gadzama told me that the proliferation of unpalatable species to the point where they virtually occlude the grazing takes as little as two years in these arid lands).

Outside the fences of the Gombole Ranch, the vegetation looked in better heart. These rangelands were grazed by the relatively small numbers of sheep and goats of the local villagers, even though they lay within the ambit of the nomadic herds of the Fulani. (There were few wild herbivores in these areas; most of the game was concentrated in the area of Damboa/Gwoza, south of Maiduguri – although occasional elephant did cross from Cameroun.) Grass growth appeared to be good (at least for that short season when there *is* growth) and because of this comparatively good ground cover the encroachment by noxious weeds seemed smaller. Most of the species present, however, were annuals. Gadzama explained that during the growing season most of the big nomadic herds moved more to the south to exploit more permanent pastures there, and thus grazing by the livestock of the resident villagers was comparatively slight. The cattle herds returned only in the dry season, by which time the annual plants had grown and seeded, and so could persist even if the standing hay was then grazed flat again by the returning herds. It certainly looked a lot better, but this system relied on very large areas so that the rotation of grazing, constrained as it was within the fenced perimeter of the official ranch, was over a much longer time period – effectively on an annual basis or thereabouts.

During the growing season the herds that remain grazed only a relatively small percentage of the grand total. The rest matures and dries but was thus still available as standing hay even after the growing season was over. As the herds grazed slowly over the huge circuit and finished the last of this standing hay, so the rains returned and the herds could begin that large migration once more. The sheer size of the area covered ensured sustainability – just as the wild herds of East Africa migrated around the Serengeti and the Masai Mara using the available vegetation in each area until the rains returned to the Serengeti plains.

But as a management system this too had its problems. It could work only if the area over which the nomadic herds wander was huge (in order to permit this natural fodder conservation in areas unoccupied in the rainy season). It worked also only if no one else was making use of the forage in these other parts of your annual circuit in the meantime, and in arid areas such as this there was also always the risk that you might lose all your dry season forage to bush fires. Finally, and especially with large herds: when animals finally did come to utilise the area, they grazed it totally flat, not moving on until every scrap was finished. While the seeds set would reclothe the land in the following rainy season, in the meantime the ground through the remainder of the dry season was bare, and thus susceptible to erosion.

The lack of winter feed was a real issue; while I was in Borno State, the loss of condition of animals through the dry season meant that many cows would fail to come into oestrus – indeed I was told that migratory Fulani cows might conceive only once in every three years, and if they did give birth, commonly could not maintain the energy cost of lactation so that calf mortality was high. Conserving or increasing the availability of dry season forage by one mechanism or another seemed essential.

One solution to this was offered by the Gujba grazing reserve we visited the following day; this was under the management of the Federal Livestock Production Unit. At Gujba, active management was undertaken of the grazing areas to improve the amount of dry season forage produce. In parallel to this, efforts were made to reduce the migrations to a system of much more local movements around a more settled base. In essence, this was a sort of settlement programme. The idea was to take over the large area currently used by 28 families of nomadic herders and to issue leases to each family for use of that land. This not only tied those families to this known area (perhaps reducing their depredations in adjacent ranges), but because they were the legal leaseholders it also reduced, in theory, the number of grazing herds using the grazing over and above those of these 28 known families. This potentially offered greater control of the numbers of people – and the size of the herds.

But the word 'potentially' is crucial, and it is hard to control the wanderings of a nomadic people who have no use for – and no respect for – government leases.

Eight
PROBLEMS WITH OVERGRAZING

With a clear work focus to concentrate upon, Gadzama and I had seemed to be getting along fine, even if there remained a reserve – and our driver pretty much kept himself to himself. Things were not helped by an artificial divide in status derived from the fact that while I was accommodated in what passed for luxury in the Maiduguri Hotel, they were quartered elsewhere. But the following day things seemed distinctly frostier. We were heading further upcountry; even driving at a steady 120 kph, we would be two days on the road. Alhaji Zanna Talha maintained his usual courteous reserve, but Gadzama was actively uncommunicative. In fact for the most part through a long and tiring day they virtually cut me out and responded only to direct questions as economically as possible. This was disquieting, because not only did it make it rather a depressing day's non-stop driving for me; it did not bode well for the work we could only complete if we worked together as a real team.

The nature of the country through which we were travelling was changing – not so much in the landscape of the arid scrublands as in the character of the settlements we passed. As we passed, I noticed differences in the styles of the individual huts, which seemed to vary consistently in the form of their thatch and in their retaining walls, which ranged from plastered mud, to constructions of loosely stacked mud bricks, progressing to others where walls – and the curtain wall surrounding the entire village – might be of woven reed in geometric patterns. The following morning over breakfast in our temporary accommodation at Potiskum, I engaged Gadzama in a long conversation: starting by discussing our work, telling him of my impressions so far, sharing with him the germs of the ideas I was beginning to see towards possible improvements. He started to thaw immensely, perhaps realising that I considered him truly an equal partner in this – indeed the senior partner, for he was far more experienced than I – and clearly both intelligent and imaginative. In passing I casually mentioned having noted the changing styles in the designs of the villages and their huts along our way, asking if they were genuine regional variation or even perhaps tribal. The thaw was total; he clearly realised that this outsider who had been sicked upon them (at his seniors' request, not his) was really interested in his country and his people, not some cold outsider come merely to teach them their own business. He became animated and chatty; I noticed later that the thaw had spread to Zanna Talha, who also started to treat me with friendship rather than simply courteous respect.

At Jakusko, they departed from the road to take me to a Fulani village – one where the people had become settled and no longer pursued their former nomadic existence. They still kept cattle as well as a few goats, but these were grazed year round in the immediate surroundings. To my amazement, I was introduced not only to the men of the village, but also to the women (albeit that they were delightfully shy). It was indeed the first and only time in my career that I had been subjected to the inspection of people who had genuinely never before seen a white skin. But these people were courteous and charming with a natural poise and nobility of manner. They showed me into their huts and brought for me a gourd of yoghourt. Remembering dire tales of such yoghourts being curdled by the additional of cow urine, I was to say the least slightly suspicious, but the stuff tasted all right and I was conscious of the honour they were according all of us as honoured and welcome guests. Perhaps I did wait to see how Gadzama and Zanna Talha managed their gourds first, before embracing mine, but …

At our destination that evening in Nguru, close to the border with Chad, I was taken into the market, where Gadzama made a point of showing me everything and explaining what I was seeing, before helping me select and buy gifts for friends and family. And for the first time, that evening we ate together. Before dinner, Gadzama was going to visit friends in the village, since he used to work here. To my delight, he invited me along too, although I was slightly apprehensive about the imposition an additional guest would place upon our host. I need not have worried. Once again, I was told, the house was *honoured* that I would wish to visit with them. A chair was brought for me (as it emerged, the only chair in the house) and I was urged to sit and was really welcomed into the heart of things. I found the openness of the people delightful and refreshing; the fact that they had so little, yet would honour an (undeserving) visitor, a little humbling.

But the ice had clearly been broken; we ate together in the rest house, and the following evening Gadzama and Zanna Talha, together with the zonal livestock officer, got together and decided that instead of all this quasi-European food I was always served, I was to eat a real Nigerian meal. The livestock officer's wife and the wives of other members of staff were called upon, and rose to the occasion with a spread of traditional Nigerian dishes, and we all sat round on the floor in the Alhaji's chalet and ate together as equals. The food, eaten with the hand, was delicious, and I felt for once truly welcomed; they for their part were delighted that I wanted to be part of it all and belong. From that time onwards we all got along famously.

So far in our travels I had noticed little evidence of actual desertification in the complete loss of vegetative cover; overgrazing by free-ranging herds had caused a transition more towards annual species of grasses and herbs, and heavier grazing still within the fenced enclosures of the ranches was apparent only really as a shortage of dry season feed and an expansion in dominance of unpalatable 'weeds' – but now, between Jakusko and Nguru, I had become aware not only of the changing architecture of the villages but of a real desertification of the range. Here again the twin systems of fenced ranches and open-range grazing sat side by side – but the effective stocking density on the open-range systems was increased because of a comparatively high human population density and because a much greater proportion of the land had thus been converted to arable.

The only solution would seem to be to enforce a reduction of livestock numbers. But this would be politically extremely difficult – to try and enforce a restriction in the numbers of animals that might be kept by the settled Kanuri or settled Fulanis – for, as with our Kenyan herders, cattle, sheep and goats were, and still are, a mark of a man's wealth and status, an economic currency. And it still wouldn't solve the problem, for it seemed Nguru was a collection point for herds over a wide surrounding area, for market (or alternatively for transport south by train), so that whatever efforts might be targeted on reducing the stocking density of resident herds, outside people would always be bringing herds to this focal point from the surrounding countryside, and from Chad and Niger. The only alternative would seem to be to increase the grazing area by taking back arable lands and establishing huge grazing reserves, but that again would seem political suicide.

This nightmare situation was clearly as much a sociological problem as a biological one. In some ways it would appear that we westerners have interfered too much by increasing population density and encouraging settlement; that the old migratory habits of the nomadic Fulani might have offered the least worst option as they followed the cycle of seasonal grasslands like wildebeest. But you can't reduce human population densities in any politically acceptable manner, and we should remember that even for the Fulanis the loss of condition of their animals through the dry season meant that many cows would fail to come into oestrus – and even for those which did conceive, calf mortality was extremely high.

It seemed indeed an intractable problem, or indeed an intractable series of problems, since here up in the north-east towards Chad the greater problems were on the open range, while back down at Gombole the problems were more to do with the management of enclosed ranches. By glorious irony, the government ranch at Nguru was a tremendous tribute to what can be achieved by good management. There were many paradoxes here, for the area was incredibly dry (with less than 8 inches of rain per year in a very restricted rainy season); yet they managed to maintain green in the grass until January. The ranch was also only some 7 square miles (compared to the much larger areas enclosed at Gombole or Gujba, in areas of significantly higher rainfall). Yet this area supported 160 cattle (all dairy) and many sheep, with no signs at all of overgrazing.

The grass sward was dense and lasted all year, with the animals more or less maintaining body condition throughout the year. The sheep indeed were employed here as browsers on the dense shrubs, to keep them in check and maintain or even improve the grazing for the cattle, but the whole offered a clear example of what could be achieved with imaginative – and intensive – management.

I was becoming more and more confused by all the apparent contradictions, and needed some time to sort out my jumbled impressions. In the end, it half-clicked into some logical frame: my presentation to the Permanent Secretary when I returned to Kano attempted to synthesise some of the problems in general with overgrazing (by livestock) on open range, as expressed by the resulting lack of dry season feed; the progressive eradication of palatable species, decreasing the carrying capacity of the range; and the encroachment of unpalatable or noxious species. These effects were in fact sequential, responses to ever-increasing grazing densities, ultimately in the extreme leading to actual desertification. It seemed to me that these changes were not a consequence of grazing pressure *per se*, but rather an interaction between animal numbers and the vegetation type (annual or perennial), productivity of soils, rainfall etc, so that a particular stocking density which produced one result in one location might produce a different result in a different area of lower rainfall. It did sort of fit – with the added complication that grazing at a certain pressure could also cause a shift from predominantly perennial vegetation to a community of annuals.

Clearly I emphasised that many of the *biological* solutions I might offer might prove difficult to implement because of the sociological or purely political considerations that were partial drivers of the problems; but I did then divide my recommendations between the very distinct systems of fenced ranches and grazing of the open range. For enclosed areas, most of which were at the time state-run, there was clearly greater potential to implement alterations to stocking density (and species), and I made clear a need to tailor livestock densities more closely to vegetation production and the length of the growing season (in relation to rainfall). It seemed to me that the success of efforts in what was otherwise the least promising area at Nguru indicated clearly how important it was to select the appropriate balance of grazing species, so that it might be appropriate in many places to replace pure monocultures of cattle by mixed grazing systems involving both cattle and sheep. Of course, given my own conviction that native ungulates with a somewhat broader dietary spectrum tend to have a lesser impact on the vegetation – something given active support by the experiences at Galana – I mentioned the possibility of ranching native ungulates as opposed to sheep and cattle. I subsequently learnt that the Ministry had already been considering the possibility of an experimental eland ranch as part of its next five-year plan.

It seemed to me that in open-range areas of unrestricted grazing the reaction of the vegetation was directly related not so much to the number of grazers overall as to the overall area over which those grazers might move in annual migrations. Where movements were more restricted – and so, if you like, the number of 'grazing days' per acre were increased – some consideration had to be given to increasing the size of area available for grazing and/or setting aside specific areas for conservation of fodder towards the dry season. In areas such as Gujba they had already had some success with trying to improve the grazing by pasture improvement and partial irrigation. If all else failed, at least some effort might be made to try and restrict the expansion of unpalatable species over critical areas by direct intervention by physical or chemical treatment. None of these ideas were new; many were already being trialled in different parts of the state. I guess my role was to try and piece it all together.

But that report was still a few days distant.

Nine

POMP AND CIRCUMSTANCE

Our return to Maiduguri coincided with a public holiday to celebrate the Salah – a commemoration of Abraham's sacrifice of a ram to God in place of his son. The Salah would be followed by a durbah, a ceremonial occasion when the emir would receive the annual reaffirmation of fealty from the various sub-chiefs who ruled under him across the state. For all that Nigeria was a modern republic, the secular power of the emirs was still great, and they took their responsibilities very seriously: hearing grievances from their people and dispensing appropriate justice within their own regions. Each still controlled an essentially feudal –even mediaeval – system of underlords and chieftains, formally committed to provide arms and fighting men in support of the emir if requested. The durbah provided ceremonial reaffirmation of that allegiance and loyalty, which was still taken very seriously and was far from having become merely colourful and historical tradition.

In the days before the Salah itself there were a number of sales of rams within the city walls. These sheep were a distinctive black and white breed, marked not unlike a Bagot goat. We saw little of Alhaji Zanna Talha during the day or two before the day of sacrifice, understanding that he wanted privacy to undertake the preparations and sacrifice his own ram for the benefit of his household.

The main part of the Salah is celebrated at home, but there was a subsidiary religious ceremony the following day in preparation for the durbah. I had been invited to attend, and waited nervously in the background as the crowds assembled in a large arena outside the town. Many arrived on donkeys, and there was a dedicated donkey park assigned outside the arena. As they amassed more and more, I noted what appeared to be pedlars working the rows of worshippers, but realised quickly that they were selling sets of prayer beads. All wore robes and, unless they sported the red fez of one who had completed the pilgrimage to Mecca, wore the entirely characteristic hats of this part of Nigeria: flat-topped cylinders intricately woven and embroidered in row upon row of vertical stripes of contrasting colours.

Alhaji Zanna Talha was by my side. It emerged that he was perhaps some relative of the emir, or a functionary within the court, for he said that the emir would be pleased if I would join him in the covered pavilion raised at one end of the arena to observe the durbah as he received the tributes from his chieftains. To say I was stunned does not describe it. Already I had seen on the outskirts massing groups of horsemen and foot soldiers in bright costumes, the horses themselves wearing heavily decorated armour and chest-plates like those of mediaeval knights at some tournament.

To be offered a more privileged view of the full ceremony was incredible fortune. I spluttered my thanks and was taken to sit under the silk canopy in the front row of the pavilion, seated directly on the emir's right hand – and with Zanna Talha to my own right hand. From this elevated position I watched each chieftain present his fighting men in turn to the emir. From the far end of the rectangular arena, mere dots in the distance, they moved slowly forward across the sand. Figures in colourful robes at the front might bear long plumed brushes with which they ceremoniously brushed a path across the dry sand towards the dais. Behind them might come rows of warriors on foot, waving the great curved scimitars of Arabia, or simply massed ranks of what seemed like families: men, women and children, walking slowly forward across the floor, although even among these groups many of the men bore a huge spear across their shoulders. All moved steadily, inexorably, forward to the steady beat of huge drums. Behind them, the horsemen, their horses resplendent in their heavily decorated breastplates and armour, the riders themselves equally glittering in gold or silver tunics or robes. All moved forward in distinct rows, and as each group reached the ground below the pavilion they would bow deeply to the emir before stepping aside to make room for the phalanx behind. Sometimes, if they were many, the horsemen might hold back a bit behind those on foot, to gallop forward at the end in a mock charge, shouting furiously and waving their scimitars above their heads, before they too bowed to the pavilion, leaving behind them a cloud of dust.

Wave after wave came forward in blues, in shimmering turquoise, in chequered harlequin patterns of reds and blues. As each chieftain had presented his subjects and armaments to the dais, they withdrew, and a slow advance would be observed from the far end of the arena of the foot soldiers and horsemen of the next chieftain. It was an amazing spectacle: the colour, the gold and silver, the pageantry of the decorated and armoured horses; perhaps each petty chief competed to outdo the sumptuousness and display of his predecessor, yet strangely it was all done with utmost seriousness and sincerity – and sheer pride. I sat, entranced, my camera bag at my feet.

Presently, the emir nudged my arm. He pointed to the camera bag; I was not taking photographs. 'No,' I replied, for while I had not been fully warned before my trip about the offence which might be caused by the exposure of too much unclothed flesh, I *had* been advised that many Muslims were unhappy about having their photographs taken, since this would produce an image of a sentient being, forbidden in Islam.

Did I wish to take photographs? – and if so, what would I use the photographs for? Hurriedly I explained that I was simply overawed by the spectacle, and if I was to take photographs it would be to show to my countrymen in admiration – indeed awe – of such a spectacle, to show what a tremendous country and a tremendous people this was. It sounds corny, yet my reply was sincere and it was not contrived for flattery; with an imperious flutter of the hand I was told to take up my camera.

It was a tremendous climax to my time in Borno, for the next day I had to fly back to Kano and present my report to officialdom – and then, after a final night in my closely guarded British Council compound, fly back home to London. I have the photographs to this day – a quite breathtaking spectacle of colour, of solemnity and of this handsome people's genuine pride in their nation and its Arab traditions.

BOOK 3

Indonesian adventure

One

ARRIVAL IN BALURAN

Baluran National Park lies on the northeast tip of the island of Java, close to the town of Banyuwangi. The area was originally designated as a hunting preserve by the Dutch colonial administration of the time, but after independence its status was changed to that of a national park. It covers an area of approximately 220 square kilometres, stretching from the coastal mangrove forests bordering the Bali Strait and rising to the central extinct volcano, Gunung Baluran, which reaches 1,247 metres at its highest point.

The vegetation of the area reflects this same gradation in altitude and the strongly seasonal pattern of rainfall. The upper and middle slopes of the volcano are covered by tropical monsoon forest, typically more evergreen than the forests of lower elevations, and with a dense understorey of bamboo and rattan. Below about 300 metres, this grades more towards a deciduous forest which covers the foothills of the volcano and the east and south-eastern parts of the park; this tends to be more open in nature and is punctuated by grassy clearings and glades. Towards the coast, areas of freshwater swamp forest are developed, which at the coast itself are replaced by a coastal fringe of mangroves. Teak has been planted on the north-western, western and south-western flanks of the Gunung Baluran mountain, reaching to the road which forms the western boundary of the park. This forms a buffer or zone reaching to 500 metres above sea level and occupying about 20 per cent of the park area.

Throughout the park are a number of areas of savannah grassland, although it is unclear whether these are natural or of anthropogenic origin. Fire has clearly played an important role in maintaining these areas of open grassland, and in a misguided attempt to prevent the fires damaging the native monsoon forest areas, the introduced shrub *Acacia nilotica* (gum Arabic) was planted around the edges of the grasslands as a fire break. Unfortunately, this species is highly invasive and has now encroached both across the savannahs themselves and deep into the forests.

For some considerable period after independence, the Dutch had maintained a research presence within Baluran; now I myself had two doctoral students working in the park. Although the individual research programmes which would be written up towards their PhD degrees were tightly focused, Simon and Martin were collaborating over much of the fieldwork to collect a range of data about the dynamics of the entire system. As one part of that, Martin was keen to attach radio-transmitter collars to a number of the little muntjak deer which lived within the park – at perhaps somewhat unexpected high density. Netting these little animals had, in the experience of others in the past, led to unacceptably high mortality, so Martin was keen to tranquillise them for capture by using blowpipe darts.

For that he wanted help, and the appropriate drugs. As his research supervisor, I was to offer both, by flying out from Southampton to offer assistance but also to carry the necessary drugs. We did in fact have impressive-looking letters of authority from the ministry in Jakarta and from the university in Southampton, but I was painfully aware that these might count for little in a crisis – and that indeed, the more impressive of them I had contrived myself. The last thing therefore that I wished to do as I arrived at long last in Jakarta was to attract any undue attention. But it had been a long flight and I was hot and sticky, as well as more than a little weary. As I reached up to the overhead locker on arrival to retrieve my overnight bag, my denim jeans caught my sticky self and ripped, leaving a long tear across my buttocks and around the crutch. To all intents and purposes my nether regions now sported less a pair of trousers than a matching pair of independent legs. I wrapped the remnants of my jeans around me as well as I could, and set off resolutely towards Immigration.

Like many frequent travellers, I do not enjoy travelling. I seem to spend an inordinate amount of my time flying (or rather: sitting around in airports *waiting* for flights or connections, for one spends a comparatively small proportion of any journey actually in the air). I am not, then an enthusiastic traveller, bearing with it stoically simply in order to arrive. I find long-haul flights themselves not threatening in any way, but intensely boring. I was not thus in the best of humours, and the wreckage of my trousers was adding insult (and anxiety) to injury. Fortunately my concern must not have shown in my face – perhaps even the damaged trousers diverted attention from too close an inspection of my luggage – for I was waved through without further ado, and Martin and I headed off to find the nearest outfitters, where I might buy a replacement set of (looser) trousers to cover my rather obvious undergarments. The tattered jeans were abandoned in a convenient street bin, before we went off to find the bus to Surabaya and thence to Banyuwangi.

Perhaps not unnaturally, I do not recall much of the bus journey – partly through simple tiredness and partly because a good deal of the time was occupied in catching up with the news from Martin. In those days before airport security and the internet, we could not communicate as easily or regularly as we might have wished: all correspondence was restricted to conventional airmail, which might take many days in passage. (Martin and I are still in close contact 20 years later, and it seems another world where we can keep in daily touch if need be by email – and I need only a single digital address, rather than having to keep a careful note of his probable itinerary and therefore to which poste restante to send any urgent communication.) I do remember being met by Simon at the park gates in their rather battered old saloon car and driven through the park to the field house the two boys occupied at Bekol. This was one of the former guest houses or hunting lodges built by the Dutch during the park's time as a hunting reserve; built largely of timber with a large veranda, it was cool and spacious, even if the table legs all had to stand in individual cans of water or kerosene to stop the forest ants or termites colonising the upper plateau and invading the computer or other more delicate equipment.

And I *certainly* remember that first evening, for I was wakened in the night by the calls of rusa all round the house and across the Bekol savannah; my arrival had coincided with the rut, and the stags had all congregated on the scrubby grassland to compete for mates and mating opportunities. The roars echoed close around the house from dusk throughout the night – there were clearly animals all around us. It was all so unexpected and so intense. For the past 25 years or so I have lived in the Highlands of Scotland, quite accustomed to the roaring of red deer stags around me each autumn, but I think I have never encountered anything quite like that first night at Bekol; there must have been 40 or more testosterone-charged males roaring all around us – and it was merely a frustration that one could not actually distinguish the animals in the gloom.

Two

BEKOL AND BAMA

Simon and Martin were good friends and had studied together as undergraduate students. They made an excellent team, for in many regards they were as chalk and cheese and thus complemented each other extremely well. Simon was perhaps the grand strategist, the man with the eye for the overview: he was a tall, well-built man in his mid-twenties, with a shock of curly brown hair and spoke always in a husky whisper; the nearest parallels I have met are always characteristic of those who have recovered from throat surgery for oesophageal cancer. In person (perhaps with the benefit of unconscious lip-reading) I was always able to cope, but over the telephone I was forever having to ask him to repeat himself. Martin was some years older, having returned to university as a mature student; disaffected of a career as an industrial chemist and by what he had experienced in the exploitative commercial world, he had become more and more interested in environmental issues. Never one to accept opinions second-hand from the more polemic 'green' organisations, he had returned to university to study more formally the actual facts beneath the rhetoric. Always patient, always incredibly good at solving practical problems, he was the steadying influence within a very amicable partnership.

Ostensibly the main projects in which the two were respectively engaged were firstly to investigate the park authorities' concerns about the possible negative effects of competition from feral domestic water buffalo within the park on the indigenous (and rare) banteng – at that time I think the world's third rarest species of wild cattle. While a domesticated (mongrelised) version still survives widely in the form of Bali cattle, at that time wild banteng were extremely endangered – persisting only in three populations within Indonesia, of which the one at Baluran was significantly the largest. In concern about negative impacts from large feral population of water buffalo which had reverted to the wild, the park authorities were already engaged in a long-term programme aimed at reducing populations through hunting and through large-scale live-capture operations (subsequently relocating the captured buffalo or selling them for redomestication). Such exercises were arguably costly (although our impression was that the local park managers were not too keen to discontinue live-capture operations, which actually generated significant personal income ...) and it was hard for the Indonesian government's Directorate of Forest Protection and Nature Conservation (PHPA) to justify the expense if the evidence for competition was not as strong as presumed. In any event, such a study would also reveal more about the behaviour and ecology of the banteng themselves, which might be of relevance to their future conservation.

In Martin's case, he was curious to understand why – when surrounded by so many predators and competitors, and faced with a dry season which lasted from May to mid-December[6] – muntjak seemed to persist within the park at such improbably high densities.[7] There could surely be relatively little food available to the muntjak over this long dry season especially when, as a comparatively primitive species with a poorly developed rumen, they are not particularly well able to digest coarse vegetation, and are reliant on a high quality diet of growing shoots and fruit.

But in truth, both banteng and muntjak were part of a much wider community of animals, including, as well as the water buffalo, rusa deer (a relative of our own red deer – and of equivalent size) and wild pigs (our own familiar wild boar, *Sus scrofa*). There were also leopards,

---

6 In Baluran the majority of the rainfall occurs during the SW monsoon, during the single wet season from December to April.

7 I use the spelling muntjak advisedly. This is not the Chinese muntjac familiar to European readers from its introduction to the United Kingdom and elsewhere, but the (related) Indian muntjak (*Muntiacus muntjak*).

and populations of the Asiatic wild dog (or dhole; in Indonesia referred to as *ajag*) – Asia's ecological counterpart of Africa's familiar hunting dogs. And this, to Simon, Martin and myself, constituted the wider challenge: so much has been done in African ecosystems to establish the 'rules' structuring communities and their interactions in tropical systems – and the rules established about the relative roles of competition, predation or active facilitation are commonly taken as a model for all such systems. But in practice the rules which structure communities might not in fact prove so universal, despite the remarkable parallel evolution of ecological roles within the two systems, so that every African species seemed to have its ecological equivalent in Asia: here was a golden opportunity to explore the full dynamics of this rich community and see if the same 'rules' did apply.

The two had established a regular programme of data collection throughout the park on a repeating cycle: collecting information on actual numbers and distribution of the different ungulate species within the area (in both wet and dry seasons) through regular patrols of fixed belt transects; collecting information opportunistically on the behaviour observed and – towards future analysis of the diets of the different species and to what extent they might overlap – making and labelling collections of fresh faeces. Because of the way that the flights had worked out, I had arrived in the middle of one such cycle of routine observation and sampling, so I elected to make myself scarce for the first week or so of my visit, to leave Simon and Martin free of any responsibility for my entertainment and allow them to return and complete their data collection for that month. While this to some extent left me kicking my heels, it also offered an opportunity for me to explore something of the park on my own and get to know parts that we might not visit during our subsequent efforts to dart muntjak.

My explorations were fortunately not to be entirely restricted to the area immediately around the house at Bekol. In addition to their elderly (and doubtless not road-legal) white saloon car, Martin and Simon, like everybody else it seemed, possessed a small motorcycle. These small 50cc machines seemed the mode of transport of preference throughout Java. Certainly, all the park staff rode these little bikes, and any trip to Banyuwangi revealed the streets to be full of them weaving their puttering paths around the horse-drawn shays, the donkey carts and cattle and the three-wheeled pedal rickshaws that passed for taxis. Astride one such diminutive machine I was free to skip up and down the (single) main access road through the park or continue on down to the coast at Bama. Abandoning my machine at any point, I could venture further into the forests or explore some of the larger savannahs.

Perhaps my favourite outing was to take the bike down to the coast and wander through the swamp forests down to the beach and the mangroves. The swamp forest itself was dark and cool, with huge trees reaching up into the canopy on ribbed and buttressed roots; the understorey of deep leaf litter full of the deeply dissected fronds of young *Corypha* palms. Gaps in the canopy allowed shafts of light to slant to the forest floor, and in these sunspots there were always lots of butterflies. There was a vague track one might follow through the forest to emerge at the far side on a white coral beach facing out into the Bali Strait, itself fringed with mangroves stretching their gnarled roots into the brackish water at the margin of the freshwater swamp forest.

The abundant butterflies were perhaps not as glamorous as this description may suggest, since they were congregating as much as anything to uncurl their delicate little probosces and sip fluids of decay from the slowly putrefying corpse of a large male banteng. Whether it had died here of natural causes or had been poached was hard to determine, although poaching was sadly not uncommon within the park. The huge, dark carcase (adult male banteng have a deep mahogany-brown – almost black – coat with a white rump patch, and strikingly white legs and feet) attracted more than just butterflies. Each time I visited I disturbed little groups of wild pigs scavenging from the carcase, which exploded away through the forest. But there were also monitor lizards, and if one didn't mind the stench (much depended on the direction of the wind at the time) and was prepared to mount patient watch, one would see them gradually returning to continue feeding on the carrion. These were not of course the dragons of Komodo, but in fairness the larger individuals were impressive enough, and it was rewarding to the patience to watch them close up, flicking their long forked tongues in the wind, hissing and squabbling over scraps of meat or access rights to a particular part of the carcase. Occasional crabs would also come in from the mangrove to filch what they could, but the monitors undoubtedly dominated the scene.

In the evenings I would return from such outings to the field house, usually to discover Simon or Martin spreading small pats of banteng and buffalo dung on home-made tinfoil platters before setting them to dry in the sun for storage. Apart from the fact they were laid to dry in the sun, it resembled nothing more than a bakehouse dedicated to chocolate confectionery, but it was vital to the study, and in due course literally tons of the stuff were transported in barrels back to the UK. I am not sure that more than the tiniest fraction was ever analysed …

Because Simon and Martin were trying to spend the maximum amount of time each day in the field, they had asked the wife of one of the park rangers to cook for them. Having finished his own meal, he would arrive each evening on *his* motorbike with the carefully packaged take-away meals for our repast. Rice is of course a staple throughout Asia, and most of the local people would primarily be eating meals of rice with some small added protein in the shape of fish or meat. But local conviction was strong that '*boulays*' (white people/Europeans) could not digest rice, and that for them to eat rice would be dangerous and possibly even fatal. Simon and Martin had tried over and over again to persuade the staff that indeed they could eat rice and liked it very well, but it had been an uphill struggle, and there was still some trace of concern on the ranger's face as he handed over each evening's tray. Clearly, the Europeans had not died *yet*, but …

Actually, the meals were extremely good; although they were primarily based around the daily ration of steamed rice, this was always accompanied by a delicious vegetable casserole, or perhaps some locally caught fish (always unrecognisable to me, always bony, but always rather good) or very occasionally chicken. Most of the materials would have had to be bought specially from the market in Banyuwangi, but fresh vegetables were not expensive in season and of course, as we were right on the coast, fresh fish was readily available. We ate very well, in fact, for comparatively little.

Our days settled for a while into a routine of early rising and breakfasting on strong black coffee. Fortunately we all took our coffee black, and all liked the strong local Javan beans. Simon was punctiliously – almost obsessively – careful about bringing the water to the boil each morning and then religiously leaving it boiling for 4–5 minutes to kill off any potential bugs before brewing the thick, dark coffee. In his defence (although I am someone who does not operate in the mornings *before* those first two cups of coffee, so that I was never-failingly impatient) while it may have been an over-precaution, it is true I never had problems with dysentery. Any scraps from the dinner of the previous evening would be hurled out the back door towards the adjoining store-room, where a fine male Javan peacock would be waiting for just this occurrence, resplendent in the iridescent green plumage of the Indonesian peafowl (as opposed to the more familiar blue of the peafowl of India). I will draw a veil over the necessary morning ablutions for, despite its European origins, the field house, like most other Indonesian houses, had a simple *mandi* which served for all lavatorial purposes, whether as a shower, for a more conventional wash or as a facility for relieving oneself.

Then we would head out in our separate directions, to meet again in the evening for our main meal of the day.

On occasions, perhaps more confident now of my fitness and realising that, after all, his geriatric supervisor would not slow him down, Simon would agree that I might accompany him on his treks out to the upper monsoon forests on the flanks of Gunung Baluran; or I would again take the little motorbike down to Bama and watch the water buffalo as they wallowed in the fast-receding pools of fresh water in the swamp forest. Often on these occasions, if I sat quietly beside the pools I would be treated to the sight of troops of leaf monkeys feeding in the trees above me: silvered langurs and long-tailed macaques. One evening, realising that while I might have become even over-familiar with the dead banteng down at the coast I had not yet actually seen a live specimen, Martin decided to take me after supper into the forest into an area where a particularly fine male was known to come of an evening. Martin having left me perched safely in a convenient crotch of a nearby tree, I seemed to wait an age: the occasional muntjak would scuttle past through the clearing in which I was stationed, their russet backs simply glimpses in the evening light, but it was nearly dusk before the banteng appeared silently in the glade ahead of me. I had heard nothing; he simply materialised and was there: the silence amazing for such a huge animal. Almost black, this individual, with the strikingly contrasting white socks and white caudal patch. I would get better views later in the visit, but it was nonetheless quite something to sit there with my first sight of this critically endangered species.

T hree

MUNTJAK AND MONKEYS

The time had come for Martin and me to focus on trying to catch some muntjak. In place of the fast-acting tranquilliser I would normally use – and because this *would* have caused real problems had I been challenged at Customs – we had prepared darts containing a cocktail which would tranquillise and relax the muntjak, but not knock them out completely. Again because of legal restrictions surrounding their importation, I had brought no firearms, for Martin was sure that if we sat quietly enough alongside regular trails, muntjak would pass in close enough proximity for us to be able to hit them with darts from improvised blowpipes. Unfortunately no one had told the muntjak.

For many days Martin and I would head into the forest within the area where he wanted to mark animals, and select a likely position close beside a trail he had often seen them use, or at the crossroads of a pair of intersecting trails. Then we would sit, and wait. Perhaps because of the season there seemed to be a constant rain of ticks from the branches above us as we sat; Martin very sensibly at least wore a hat, but these tiny nymphal seed ticks got into every crevice. We saw plenty of muntjak, but they saw us too – and even those which did not trotted so rapidly along the familiar trails that there was never time to take a shot. In many days, never once did one stop long enough within range for us to release a dart. And every evening, we would return home to strip down in the mandi and sluice off thousands of those wretched little ticks … It grew to be extremely frustrating, and in truth a little boring – perhaps we were too focused on waiting for muntjak, but to my memory there seemed very little other wildlife around to distract one from the ticks.

But the muntjak *were* a puzzle. These little ungulates are what ecologists call 'concentrate-selectors'. Because they do not have the intestinal capacity to ferment and digest large quantities of vegetation, they are reliant on selecting the tips of branches, growing shoots and leaves, flowers and fruits – which was fine during the growing season of the (short) rainy season between December and April, but left them pretty scant pickings for the remaining eight months of the year.

Further, while the larger species such as buffalo, banteng or rusa were chiefly grazers within the park's savannahs, the wild pigs also favoured the fruits and more 'concentrated' foodstuffs on which the muntjak were dependent. Pigs aren't fussy either: happy to take banteng or other carrion when it presented, they were also quite capable of killing and eating muntjak fawns while they were still vulnerable in the first few days of life. How come muntjak were present in the park in densities higher than we had ever observed elsewhere? And (when finally Martin did manage to get some radio-collars on a few individuals, after my own departure) why didn't they appear to leave the park for richer areas during the long dry season?

For the first, muntjak of all species are highly productive. They reach puberty early and, with no seasonality of breeding apart from in response to the rains, have a high reproductive output. As long as food availability during the rainy season is good (and sufficient to maintain lactation) a few losses to pigs are probably not significant. While muntjak are also a favoured prey of the dhole, densities of dogs in the park at that juncture were comparatively low (and their preferred habit, like that of their African cousins, is to stay in any given area only for a short period before moving on again to terrorise the neighbourhood somewhere else within their extensive home range for a while, before moving on again).

The curiosity, then, was how the high muntjak populations survived in the dry season, when there was little fruit available and such vegetation as was within their reach was over-mature and sere. The answer quite simply, lies with the monkeys.

Although muntjak are not arboreal, in the dry season they became arboreal by proxy. They haunted the various troops of silvered langurs and long-tailed macaques, following them through the forest and picking up the fresh detritus they littered beneath their feeding trees. Monkeys are messy feeders: they grab whole bunches of leaves or flowers or fruit, biting off a few snatches, before discarding the rest, which drops ultimately to the forest floor. This sort of facilitation between ungulates and monkeys had been observed before, in India; in this case, the langurs gain from the chital because as these elegant deer graze their way through the forest, their hooves disturb insects and small reptiles which the langurs are quick to exploit; the chital gain in this mutualistic relationship because the langurs are always alert for predators, and their presence and keen eyesight increases the chances that any lurking predator will be detected well in advance, losing the element of surprise.

But while in Baluran there would appear to be a close relationship between muntjak and feeding monkeys, there is little evidence to suggest that the monkeys gain anything from the presence of the muntjak. Indeed, long-tailed macaques, which unlike the silvered langurs feed on the ground as well as in the canopy, actively chase muntjak away from dropped fruits. Perhaps the real clue to the unusually high densities maintained year round is the presence of the introduced *Acacia nilotica*. Following a series of serious bush fires which devastated parts of the park, this species of acacia was introduced to Baluran in 1968, planted around the edges of the main savannah areas within the park with the intention of establishing fire breaks which would resist brush fires and prevent fire spreading from open areas into the surrounding woodlands.

But as so often with well-intentioned introductions, things did not quite go as planned. The *Acacia nilotica* proved extraordinarily invasive and spread widely, colonising a significant proportion of what had previously been open woodland or open savannah, which led in turn to a much-reduced area of grassland, and with any open areas remaining dominated by a ground flora of relatively unpalatable forbs in place of annual grasses. Such conditions are not ideal for the rusa deer and water buffalo which predominantly feed on grass, but provide muntjak with increased areas of abundant food (and cover in the wet season). Tree legume seed pods are eaten by muntjak and the other ungulates as soon as they are available (mid-dry season onwards), and *Acacia nilotica* seed pods appear to be a major foodstuff for animals which can access them. This species has a wide flowering and fruiting period, and as it spread rapidly through the park, it is probable that it began to provide a critical dry season food resource.

One might imagine that this, like the leaves and fruits so helpfully discarded by feeding monkeys, might be a resource that is not freely available to muntjak because of their limited browsing height. However two factors act to increase their availability in practice: the larger ungulates break down trees to access the seed pods, and local people harvest the seed pods and in the process leave seed pods on the ground. And there is no question that the muntjak make good use of this resource. Like other small ungulates, muntjak regurgitate indigestible food particles, or those too large to pass easily through the gut, and piles of clean *A. nilotica* seeds were commonly encountered in the park, which are likely to be from muntjak because the seeds are small enough to pass intact through the gut of banteng and the other larger species. Here again then, the muntjak are dependent on the foraging activities of the other animals in their environment to give them access to critical dry season foods, and it would appear that it is all these various interactions which contribute to their ability to maintain such unusually high densities in Baluran National Park throughout the seasons.

As for the impact of water buffalo on banteng, Simon's studies suggested that there was little competition for food, certainly at the current densities to which populations had already been reduced (since by definition, competition can only occur if resources are in some way limited in abundance, and there seemed more than sufficient food for both species at all times of year). Indeed, his researches revealed that while current populations of water buffalo might have been reinforced by escapes of domestic animals, there was quite some evidence to suggest that buffalo were native to the area; in such a case, after a long period of coexistence, it was likely that co-evolution would have ensured that any potential for competition was minimal. There was evidence, however, of some competition for access to clean water over the dry season; the buffaloes' habit of wallowing in any available pools stirred up the mud and tainted the pools, and thus made access to clean water difficult for banteng – as well as for rusa, or muntjak, come to that. The only fresh water became restricted to the upper slopes of Gunung Baluran itself, where the thinner brown soils appeared to be unsuitable for making wallows in this way, probably because they were more free-draining.

To be honest, the larger problem threatening banteng at the time seemed to be a significant amount of poaching within the park, whether illegal or sanctioned to some degree in sympathy for the needs of the local population of the surrounding area. The carcase I had stumbled upon at Bama was neither the first nor the last we were to encounter, and an increasing amount of Simon and Martin's time day to day came to be spent in patrolling the park itself and the removal of illegal snares. Wild dogs would provide an additional pressure on the banteng.

Four

A DANGEROUS DILEMMA

The local people firmly believe that there are two species of ajag: the hunting packs of the Asiatic wild dog and the little grey ajag that live in the lavatories; these are apparently significantly smaller, but they also hunt in packs and if they bite you they will kill you because of their poisoned teeth. I never encountered either, for the dhole were uncommon in Baluran at the time of my visit. But they arrived in force some time later. These innocent killers, as Hugo von Lawick dubbed their African cousins, were well able to take even adult banteng cows (if not a match for the larger bulls), and certainly made inroads into the population of young calves and juveniles

In a fully connected natural system, the dogs might have stayed for some few days and moved on (as we ourselves would later) to, for example, Alas Purwo. For when prey becomes scarce it becomes harder and harder to catch: the law of diminishing returns dictates that it is inefficient to stay on when it takes increasing effort to catch the few remaining individuals, and the dogs' departure then allows the prey population to recover before the next invasion of the red devil.

But over the years the extensive felling of forest to clear ground for agriculture or the planting of teak or oil palms has resulted in the landscape becoming progressively fragmented; the few remaining areas of natural habitat have persisted as isolated islands within a sea of managed land densely populated by humans and their activities. Connectivity has been lost, and the freedom of the dogs to disperse to other areas through this man-managed landscape severely constrained. Sure, they *could* come and go (and would in the end) but such onward movement was to an extent inhibited by the need to cross the human-dominated hinterland; in addition, at least in Baluran the high density of muntjak provided a reservoir of fall-back prey which they could rely upon as banteng became more scarce – allowing them to persist in the area and inevitably maintain a continued pressure on the banteng, even if they had become more reliant on the abundant muntjak.

It was a worrying situation, and something of a conservational dilemma. Here – because of the fragmentation of natural habitat within the wider area – was a closed arena in which one of the largest populations of the world's third-rarest species of wild cattle (and amongst the top ten of the world's most endangered large herbivores overall) was being directly threatened by one of the world's rarest (and highly protected) carnivores. Should one intervene (and if so how)? Or should one simply leave the drama to play to its conclusion (even knowing that the tragedy was a direct result of human intervention in the first place). Letters rushed to and fro – to the PHPA (in charge of Indonesia's protected lands) and to IUCN in Switzerland. No one, it seemed, wanted to be the person to make the decision (and take the buck). In the end, the ajag moved on – but the banteng population had taken a real knock, and this is a problem which will doubtless recur.

The other main population of banteng left in eastern Java was some miles to the south, at Alas Purwo. And while the main part of Simon's work was focused on a more detailed study of the interaction between banteng and buffalo in Baluran (such delightful alliteration) he also made regular visits to this second reserve to estimate population size and assess the age structure within that population as a way of assessing its overall productivity. We drove down the coast (it was only after my return to the UK that I discovered that at that time neither Martin or Simon actually possessed a driving licence), the journey reinforcing that the entire area between Baluran and Alas Purwo itself was a densely populated and intensely man-managed landscape of plantations, villages and crop fields.

Occupying the whole of the remote Blambangan peninsula on the south-eastern tip of Java, Alas Purwo, even then, was better known for its spectacular beaches and its surf: a paradise for surfers.

We drove through the park entrance across the narrow neck of the peninsula, and walked out through the trees onto a wide beach of brilliant coral white, sweeping in a wide arc around an empty bay. Waving palms, dazzling sands and the sun glittering from the intensely blue waters of the Indian Ocean – it looked for all the world like a travel poster actually come to life with no need for airbrushing. And apart from ourselves: empty.

We were staying overnight in one of the rest houses within the park, which, like the field house at Bekol, looked out over a stretch of savannah against the edge of the forest – only in this case a savannah without the dreaded *Acacia nilotica*. In the evening light a huge herd of banteng grazed peacefully across the grassland – scores of them, such as I had never seen in Baluran. Half a dozen adult bulls, black and massive with their white stockings, moved amongst the mixed herd; among them scores of cows, many accompanied by young calves. In sharp contrast to the massive blackness of the bulls, banteng females are a delicate Jersey-cow fawn (similar to what is believed to have been the case for the European aurochs, with their massive black males and smaller cows the colour of clotted cream), so that to the eye the clearing seemed full of banteng of all shapes and shades. It was a fantastic spectacle; we counted and recounted to get numbers, and age and sex structure, but in truth it was just a simple pleasure to watch the animals as they moved calmly across the clearing. In the morning they had gone. Some few weeks later, ajag entered the park.

Poaching continued unabated within Baluran. Every day we would encounter groups of people entering the park to collect rattan, or bird-catchers with the bird-lime set to catch the park's smaller birds as they came to the ground in search of insects or other scraps. There was equally no question that banteng and rusa were being shot, and every day would provide another haul of illegal snares. Increasingly it became apparent that whatever the academic interests of the place, conservational concerns were enormous – and if Martin and Simon did not take an active part in trying to control the poaching they would have little to study. At the very least there would be little point in their studies directed towards improving management strategies for the park's large herbivores, if in the meantime those were harried to extinction. Like Dian Fossey before them in Rwanda, an increasing proportion of their time was thus spent in patrolling, looking for snares, in meeting with officials in the park headquarters or further afield at the PHPA offices in Jakarta, trying to instigate some more official action. Both did complete their fieldwork, but Simon never formally wrote up the results of his studies since he had become increasingly engaged in the political battles to try and save the park and its wildlife. At the time of writing, both Martin and Simon still work full-time in international conservation, both still in Asia. For my part, my travels were over. I packed my bags, now mercifully free of narcotic (if veterinary) drugs and flew back home.

AFTERWORD

An enthusiastic naturalist as a child (and one who was always slightly nervous of making a career out of my passion in case it should become routine and rob me of some of the joy of my hobby), I was also an avid consumer of travel books – whether or not these related to natural history. I read of the expeditions of early colonial explorers, I read with envy David Attenborough's stories of his Zoo Quest adventures and greedily savoured Gerald Durrell's accounts of his animal-collecting trips across the world. Somehow I never expected to be offered similar opportunities to travel to exotic destinations in the course of my own work as a biologist.

I consider myself extremely fortunate, indeed privileged, that in travelling as a professional one is not simply a casual visitor, afforded perhaps a rather superficial touristic view of a country and its culture. Of course, like any enthusiastic naturalist, I enjoy the wildlife element: experiencing habitats and vegetation systems new to me, seeing wonderful and exotic species on my travels. But that enjoyment is increased enormously by the focus of having a 'mission', a job to do. Because my visits have a clear objective and a clear purpose, I become implicitly more 'engaged'. The understanding required, if you are to solve problems and resolve issues, requires that you delve deeper and become in some way much more intimately involved; that fuller involvement somehow sharpens one's appreciation. At the same time, because one has a job to do, often working alongside local counterparts, one gains a different perspective of the country and its culture than might be apparent to a more casual "outside" observer; to some extent at least, it lets one inside.

The stories in these pages may seem, to some, disconnected. After all, they are chosen from among a working lifetime of such trips. But in fact they do have a theme; all of them reflect the realisation that with a job of work to do, especially working with wildlife, one gains a personal involvement which opens a new window on each new country visited. This book is not simply a travelogue; that would have little value – others have visited all the places I describe here. And while I worry that by going into detail of some of the hard science underlying my biological studies I may risk boring some of my readers, it is that exploration of the biological dynamics which opens for me that unique window. And hopefully some of those same biological insights will have been of interest to you, too.

Rory Putman, Banavie 2019

Other books for your interest

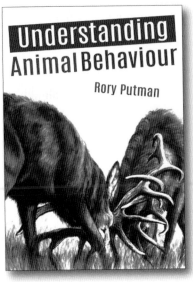

'... he brings the subject alive in this engaging book'. *ECOS*

'This is a very readable book... This book comes highly recommended and will be valuable to anyone with an interest in animal behaviour, be it academic or casual, adding a new insight to a fascinating subject'. *Deer Journal*

£18.99   978-184995-330-6

'I found the book thought-provoking with excellent examples of why we need to look at animal welfare from the point of view of the animals'. Alan Wright, *Lancashire Wildlife Trust*

£22.50   978-184995-366-5

'...What a fascinating book. ...describes his adventures (and misadventures) in a variety of far flung places as diverse as the Saudi Desert and the Niger River... The stories are beautifully told and demonstrate not only an incredible knowledge of wildlife world-wide but of the influence, good or bad, of humans on some species and their habitat. ... Gone Wild is a world tour for wildlife enthusiasts and is a book I would thoroughly recommend'. *Wildlife Detective, The blog of Alan Stewart*

£16.99   978-184995-177-7

'...his account of his journey from the bird room of the Natural HIstory Museum to these dangerous swamplands - and many other wildlife havens around the globe - is a hugely enjoyable read'. *BBC Wildlife*

£18.99   978-184995-009-1

Whittles Publishing, Dunbeath, Caithness, Scotland, KW6 6EG
Tel: 01593 731333; *info@whittlespublishing.com; www.whittlespublishing.com*

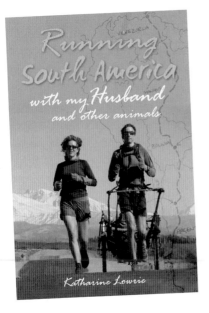